CAMBRIDGE MATHEMATICAL SERIES

CONIC SECTIONS

GEORGE BELL & SONS

LONDON: YORK STREET, COVENT GARDEN
AND NEW YORK, 66, FIFTH AVENUE
CAMBRIDGE: DEIGHTON, BELL & CO.

CONIC SECTIONS

TREATED GEOMETRICALLY

BY

W. H. BESANT Sc.D. F.R.S.
FELLOW OF ST JOHN'S COLLEGE CAMBRIDGE

NINTH EDITION REVISED AND ENLARGED

LONDON
GEORGE BELL AND SONS
1895

Cambridge:

PRINTED BY J. & C. F. CLAY,

AT THE UNIVERSITY PRESS.

PREFACE TO THE FIRST EDITION.

In the present Treatise the Conic Sections are defined with reference to a focus and directrix, and I have endeavoured to place before the student the most important properties of those curves, deduced, as closely as possible, from the definition.

The construction which is given in the first Chapter for the determination of points in a conic section possesses several advantages; in particular, it leads at once to the constancy of the ratio of the square on the ordinate to the rectangle under its distances from the vertices; and, again, in the case of the hyperbola, the directions of the asymptotes follow immediately from the construction. In several cases the methods employed are the same as those of Wallace, in the Treatise on Conic Sections, published in the *Encyclopaedia Metropolitana.*

The deduction of the properties of these curves from their definition as the sections of a cone, seems *à priori* to be the natural method of dealing with the subject, but experience appears to have shewn that the discussion of conics as defined by their plane properties is the most suitable method of commencing an elementary treatise, and accordingly I follow the fashion of the time in taking that order for the treatment of the subject. In Hamilton's book on *Conic Sections*, published in the middle of the last century, the properties of the cone are first considered, and the advantage of this method of commencing the subject, if the use of solid figures be not objected to, is especially shewn in the very general theorem of Art. (156). I have made much use of this treatise, and, in fact, it contains most of the theorems and problems which are now regarded as classical propositions in the theory of Conic Sections.

I have considered first, in Chapter I., a few simple properties of conics, and have then proceeded to the particular properties of each curve, commencing with the parabola as, in some respects, the simplest form of a conic section.

It is then shewn, in Chapter VI., that the sections of a cone by a plane produce the several curves in question, and lead at once to their definition as loci, and to several of their most important properties.

A chapter is devoted to the method of orthogonal projection, and another to the harmonic properties of curves, and to the relations of poles and polars, including the theory of reciprocal polars for the particular case in which the circle is employed as the auxiliary curve.

For the more general methods of projections, of reciprocation, and of anharmonic properties, the student will consult the treatises of Chasles, Poncelet, Salmon, Townsend, Ferrers, Whitworth, and others, who have recently developed, with so much fulness, the methods of modern Geometry.

I have to express my thanks to Mr R. B. Worthington, of St John's College, and of the Indian Civil Service, for valuable assistance in the constructions of Chapter XI., and also to Mr E. Hill, Fellow of St John's College, for his kindness in looking over the latter half of the proof-sheets.

I venture to hope that the methods adopted in this treatise will give a clear view of the properties of Conic Sections, and that the numerous Examples appended to the various Chapters will be useful as an exercise to the student for the further extension of his conceptions of these curves.

W. H. BESANT.

CAMBRIDGE,
March, 1869.

PREFACE TO THE NINTH EDITION.

In the preparation of this edition I have made many alterations and many additions. In particular, I have placed the articles on Reciprocal Polars in a separate chapter, with considerable expansions. I have also inserted a new chapter, on Conical Projections, dealing however only with real projections.

The first nine chapters, with the first set of miscellaneous problems, now constitute the elementary portions of the subject. The subsequent chapters may be regarded as belonging to higher regions of thought.

I venture to hope that this re-arrangement will make it easier for the beginner to master the elements of the subject, and to obtain clear views of the methods of geometry as applied to the conic sections.

A new edition, the fourth, of the book of solutions of the examples and problems has been prepared, and is being issued with this new edition of the treatise, with which it is in exact accordance.

W. H. BESANT.

December 14, 1894.

CONTENTS.

CHAPTER VII.

CHAPTER VIII.

CHAPTER IX.

CHAPTER X.

CHAPTER XI.

CHAPTER XII.

CHAPTER XIII.

CHAPTER XIV.

CHAPTER XV.

CONIC SECTIONS.

INTRODUCTION.

DEFINITION.

If a straight line and a point be given in position in a plane, and if a point move in a plane in such a manner that its distance from the given point always bears the same ratio to its distance from the given line, the curve traced out by the moving point is called a Conic Section.

The fixed point is called the Focus, and the fixed line the Directrix of the conic section.

When the ratio is one of equality, the curve is called a Parabola.

When the ratio is one of less inequality, the curve is called an Ellipse.

When the ratio is one of greater inequality, the curve is called an Hyperbola.

These curves are called Conic Sections, because they can all be obtained from the intersections of a Cone by planes in different directions, a fact which will be proved hereafter.

It may be mentioned that a circle is a particular case of an ellipse, that two straight lines constitute a particular case of an hyperbola, and that a parabola may be looked upon as the limiting form of an ellipse or an hyperbola, under certain conditions of variation in the lines and magnitudes upon which those curves depend for their form.

The object of the following pages is to discuss the general forms and characters of these curves, and to determine their most important properties by help of the methods and relations developed in the first six books, and in the eleventh book of Euclid, and it will be found that, for this purpose, a knowledge of Euclid's Geometry is all that is necessary.

The series of demonstrations will shew the characters and properties which the curves possess in common, and also the special characteristics wherein they differ from each other; and the continuity with which the curves pass into each other will appear from the definition of a conic section as a Locus, or curve traced out by a moving point, as well as from the fact that they are deducible from the intersections of a cone by a succession of planes.

CHAPTER I.

PROPOSITION I.

The Construction of a Conic Section.

1. Take S as the focus, and from S draw SX at right angles to the directrix, and intersecting it in the point X.

DEFINITION. *This line SX, produced both ways, is called the Axis of the Conic Section.*

In SX take a point A such that the ratio of SA to AX is equal to the given ratio; then A is a point in the curve.

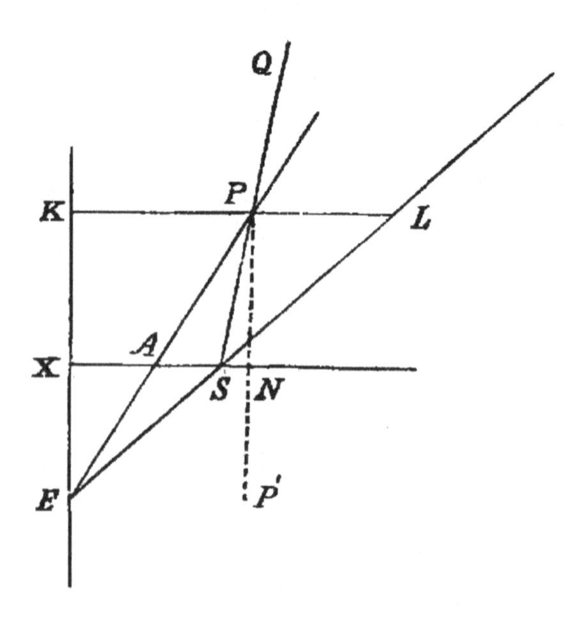

DEF. *The point A is called the Vertex of the curve.*

In the directrix EX take any point E, join EA, and ES, produce these lines, and through S draw the straight line SQ making with ES produced the same angle which ES produced makes with the axis SN.

Let P be the point of intersection of SQ and EA produced, and through P draw LPK parallel to NX, and intersecting ES produced in L, and the directrix in K.

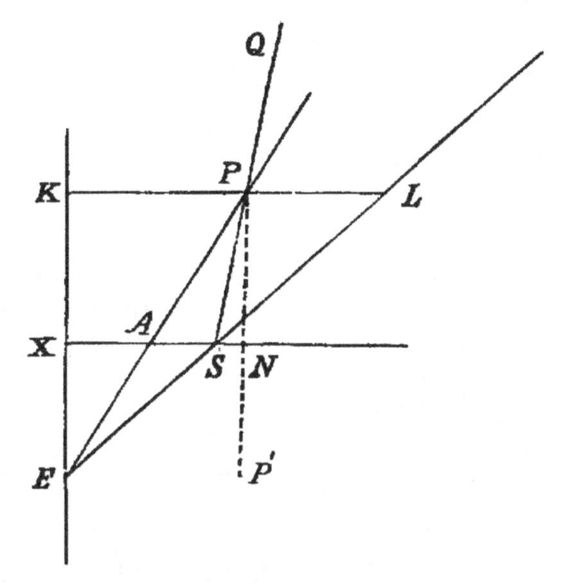

Then the angle PLS is equal to the angle LSN and therefore to PSL;

Hence
$$SP = PL.$$

Also
$$PL : AS :: EP : EA$$
$$:: PK : AX;$$
$$\therefore PL : PK :: AS : AX;$$
and
$$\therefore SP : PK :: AS : AX.$$

The point P is therefore a point in the curve required, and by taking for E successive positions along the directrix we shall, by this construction, obtain a succession of points in the curve.

If E be taken on the upper side of the axis at the same distance from X, it is easy to see that a point P will be obtained below the axis, which will be similarly situated with regard to the focus and directrix. Hence it follows that the axis divides the curve into two similar and equal portions.

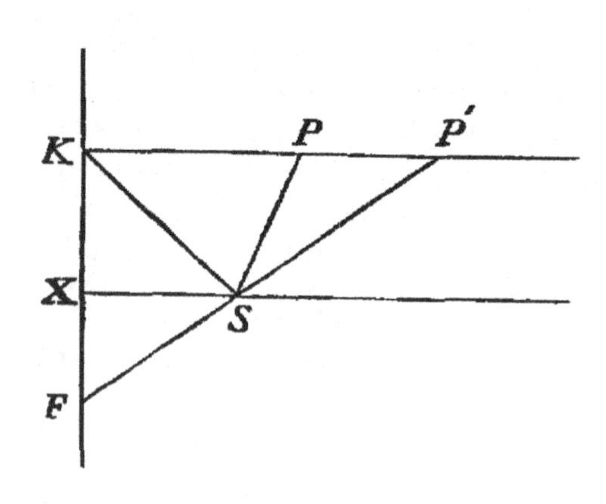

Another point of the curve, lying in the straight line KP, can be found in the following manner.

Through S draw the straight line FS making the angle FSK equal to KSP, and let FS produced meet KP produced in P'.

Then, since KS bisects the angle PSF,

$$SP' : SP :: P'K : PK;$$

$$\therefore SP' : P'K :: SP : PK,$$

and P' is a point in the curve.

2. DEF. *The Eccentricity. The constant ratio of the distance from the focus of any point in a conic section to its distance from the directrix is called the eccentricity of the conic section.*

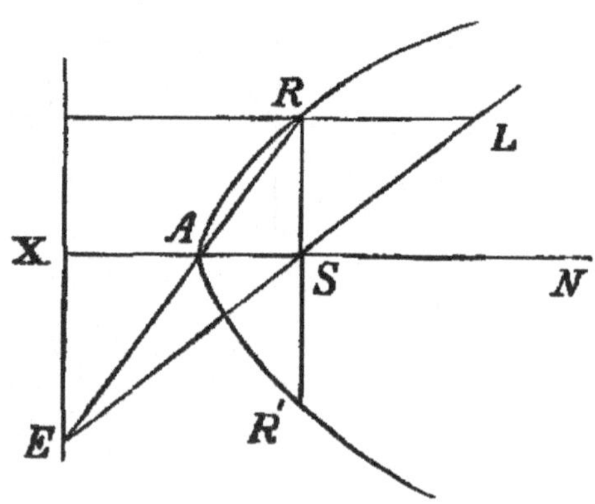

The Latus Rectum. If E be so taken that EX is equal to SX, the angle PSN, which is double the angle LSN, and therefore double the angle ESX, is a right angle.

For, since $EX = SX$, the angle $ESX = SEX$, and, the angle SXE being a right angle, the sum of the two angles SEX, ESX, which is equal to twice ESX, is also equal to a right angle.

Calling R the position of P in this case, produce RS to R', so that $R'S = RS$; then R' is also a point in the curve.

DEF. *The straight line RSR' drawn through the focus at right angles to the axis, and intersecting the curve in R, and R', is called the Latus Rectum.*

It is hence evident that the form of a conic section is determined by its eccentricity, and that its magnitude is determined by the magnitude of the latus rectum, which is given by the relation

$$SR : SX :: SA : AX.$$

3. Def. *The straight line PN (Fig. Art. 1), drawn from any point P of the curve at right angles to the axis, and intersecting the axis in N, is called the Ordinate of the point P.*

If the line PN be produced to P' so that $NP' = NP$, the line PNP' is a *double ordinate* of the curve.

The latus rectum is therefore the double ordinate passing through the focus.

Def. *The distance AN of the foot of the ordinate from the vertex is called the Abscissa of the point P.*

Def. *The distance SP is called the focal distance of the point P.*

It is also described as the radius vector drawn from the focus.

4. We have now given a general method of constructing a conic section, and we have explained the nomenclature which is usually employed. We proceed to demonstrate a few of the properties which are common to all the conic sections.

For the future the word conic will be employed as an abbreviation for conic section.

Prop. II. *If the straight line joining two points P, P' of a conic meet the directrix in F, the straight line FS will bisect the angle between PS and $P'S$ produced.*

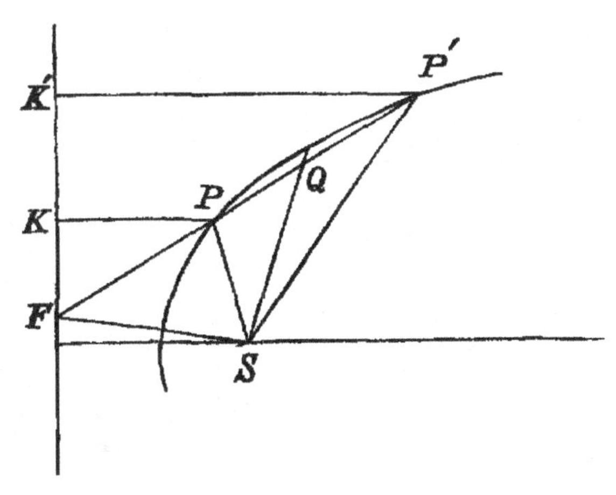

Draw the perpendiculars PK, $P'K'$ on the directrix.

Then $$SP : SP' :: PK : P'K'$$
$$:: PF : P'F.$$

Therefore FS bisects the outer angle, at S, of the triangle PSP'. (Euclid VI., A.)

Cor. *If SQ bisect the angle PSP', it follows that FSQ is a right angle.*

5. Prop. III. *No straight line can meet a conic in more than two points.*

Employing the figure of Art. 4, let P be a point of the curve, and draw any straight line FP.

Join SF, draw SQ at right angles to SF, and SP' making the angle QSP' equal to QSP; then P' is a point of the curve.

For, since SF bisects the outer angle at S,

$$SP' : SP :: P'F : PF,$$
$$:: P'K' : PK$$

or
$$SP' : P'K' :: SP : PK,$$

and therefore, P' is a point of the curve, also, there is no other point of the curve in the straight line FPP'.

For suppose if possible P'' to be another point; then, as in Article (4), SQ bisects the angle PSP''; but SQ bisects the angle PSP'; therefore P'' and P' are coincident.

6. Prop. IV. *If QSQ' be a focal chord of a conic, and P any point of the conic, and if QP, $Q'P$ meet the directrix in E and F, the angle ESF is a right angle.*

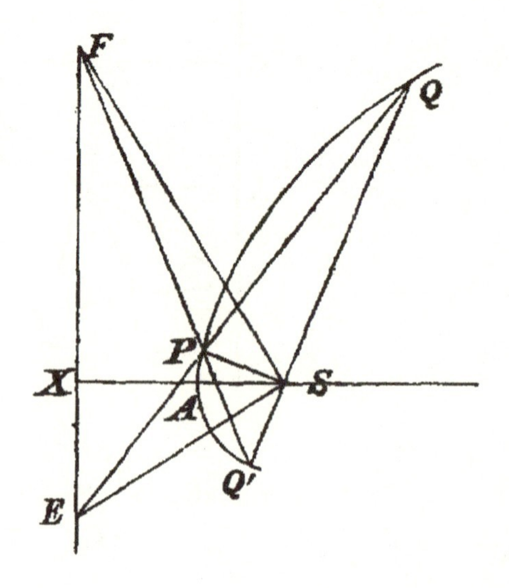

For, by Prop. II., SE bisects the angle PSQ', and SF bisects the angle PSQ;

hence it follows that ESF is a right angle.

This theorem will be subsequently utilised in the case in which the focal chord $Q'SQ$ is coincident with the axis of the conic.

7. Prop. V. *The straight lines joining the extremities of two focal chords intersect in the directrix.*

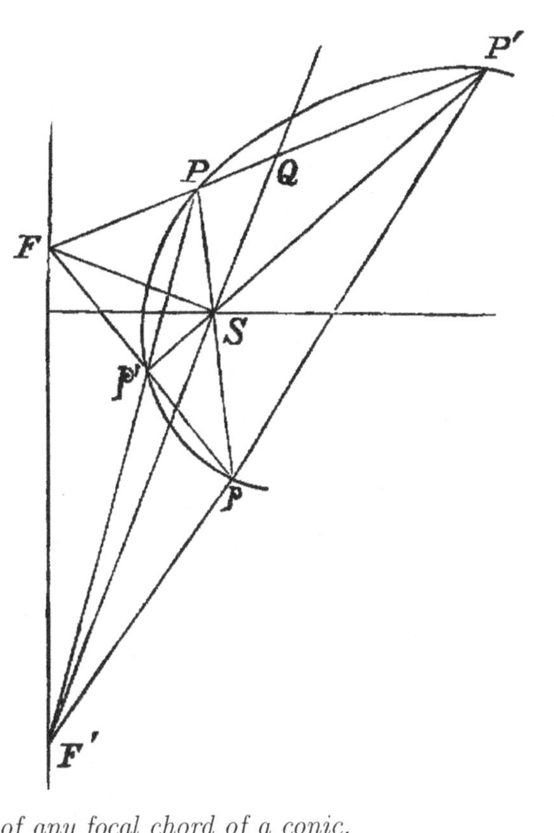

If PSp, $P'Sp'$ be the two chords, the point in which PP' meets the directrix is obtained by bisecting the angle PSP' and drawing SF at right angles to the bisecting line SQ. But this line also bisects the angle pSp'; therefore pp' also passes through F.

The line SF bisects the angle PSp', and similarly, if QS produced, bisecting the angle pSp', meet the directrix in F', the two lines Pp', $P'p$ will meet in F'. It is obvious that the angle FSF' is a right angle.

8. Prop. VI. *The semi-latus rectum is the harmonic mean between the two segments of any focal chord of a conic.*

Let PSP' be a focal chord, and draw the ordinates PN, $P'N'$.

Then, the triangles SPN, $SP'N'$ being similar,

$$SP : SP' :: SN : SN'$$
$$:: NX - SX : SX - N'X$$
$$:: SP - SR : SR - SP',$$

since SP, SR, SP' are proportional to NX, SX, and $N'X$.

Cor. Since $SP : SP - SR :: SP \,.\, SP' : SP \,.\, SP' - SR \,.\, SP'$, and $SP' : SR - SP' :: SP \,.\, SP' : SR \,.\, SP - SP \,.\, SP'$, it follows that

$$SP \,.\, SP' - SR \,.\, SP' = SR \,.\, SP - SP \,.\, SP';$$

$$\therefore \quad SR \,.\, PP' = 2SP \,.\, SP'.$$

Hence, if PSP', QSQ' are two focal chords,

$$PP' : QQ' :: SP \,.\, SP' : SQ \,.\, SQ'.$$

9. PROP. VII. *A focal chord is divided harmonically at the focus and the point where it meets the directrix.*

Let PSP' produced meet the directrix in F, and draw PK, $P'K'$ perpendicular to the directrix, fig. Art. 8.

Then $PF : P'F :: PK : P'K'$

$$:: SP : SP'$$

$$:: PF - SF : SF - P'F;$$

that is, PF, SF, and $P'F$ are in harmonic progression, and the line PP' is divided harmonically at S and F.

10. *Definition of the Tangent to a curve.*

If a straight line, drawn through a point P of a curve, meet the curve again in P', and if the straight line be turned round the point P until the point P' approaches indefinitely near to P, the ultimate position of the straight line is the tangent to the curve at P.

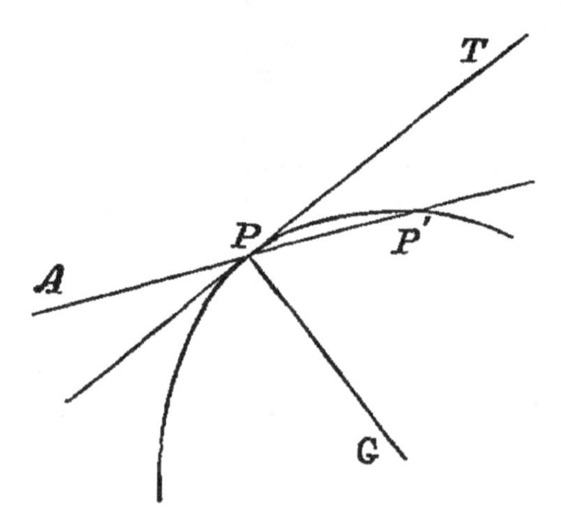

Thus, if the straight line APP' turn round P until the points P and P' coincide, the line in its ultimate position PT is the tangent at P.

DEF. *The normal at any point of a curve is the straight line drawn through the point at right angles to the tangent at that point.*

Thus, in the figure, PG is the normal at P.

Prop. VIII. *The straight line, drawn from the focus to the point in which the tangent meets the directrix, is at right angles to the straight line drawn from the focus to the point of contact.*

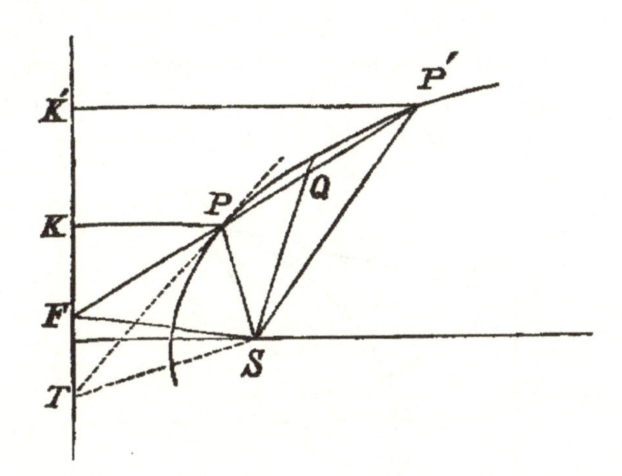

It is proved in Art. (4) that, if FPP' is a chord, and if SQ bisects the angle PSP', FSQ is a right angle.

Let the point P' move along the curve towards P; then, as P' approaches to coincidence with P, the straight line FPP' approximates to, and ultimately becomes, the tangent TP at P.

But when P' coincides with P, the line SQ coincides with SP, and the angle FSP, which is ultimately TSP, becomes a right angle.

Or, in other words, the portion of the tangent, intercepted between the point of contact and the directrix, subtends a right angle at the focus.

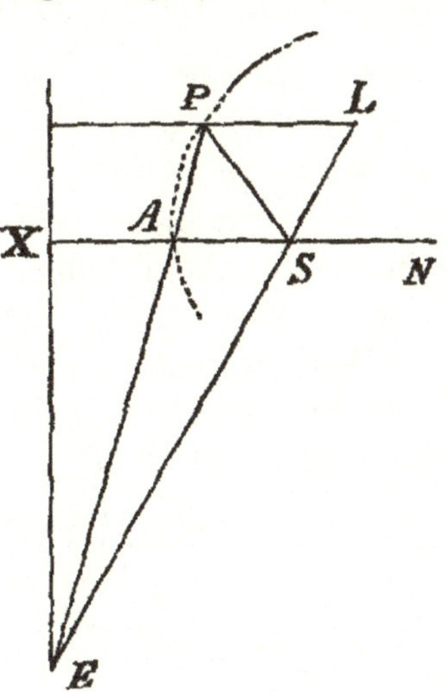

11. Prop. IX. *The tangent at the vertex is perpendicular to the axis.*

If a chord EAP be drawn through the vertex, and the point P be near the vertex, the angle PSA is small, and LSN, which is half the angle PSN, is nearly a right angle.

Hence it follows that when P approaches to coincidence with A, the point E moves off to an infinite distance and the line EAP, which is ultimately the tangent at A, becomes parallel to LSE, and is therefore perpendicular to AX.

12. PROP. X. *The tangents at the ends of a focal chord intersect on the directrix.*

For the line SF, perpendicular to SP, meets the directrix in the same point as the tangent at P; and, since SF is also at right angles to SP', the tangent at P' meets the directrix in the same point F.

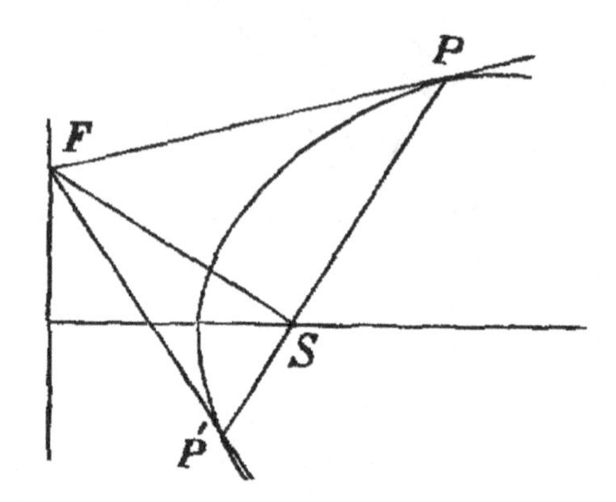

Conversely, if from any point F in the directrix tangents be drawn, the chord of contact, that is, the straight line joining the points of contact, will pass through the focus and will be at right angles to SF.

COR. Hence it follows that the tangents at the ends of the latus rectum pass through the foot of the directrix.

13. PROP. XI. *If a chord $P'P$ meet the directrix in F, and if the line bisecting the PSP' meet the curve in q and q', Fq and Fq' will be the tangents at q and q'.*

Taking the figure of Art. 7, the line SQ meets the curve in q and q', and, since SF is at right angles to SQ, it follows, from Art. 12, that Fq and Fq' are tangents.

Hence if from a point F in the directrix tangents be drawn, and also any straight line FPP' cutting the curve in P and P', the chord of contact will bisect the angle PSP'.

14. PROP. XII. *If the tangent at any point P of a conic intersect the directrix in F, and the latus rectum produced in D,*

$$SD : SF :: SA : AX.$$

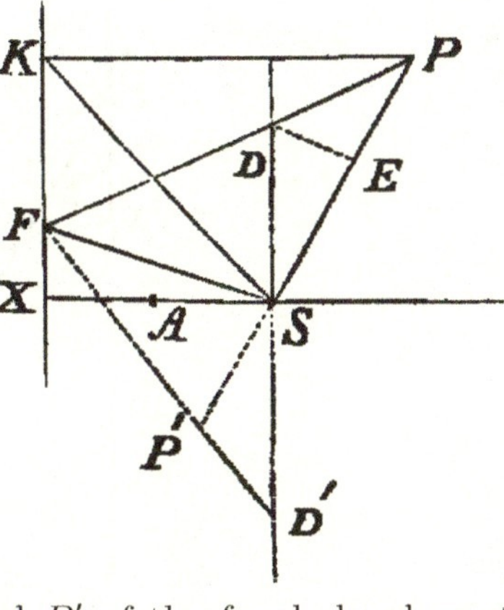

Join SK; then, observing that FSP and FKP are right angles, a circle can be described about $FSPK$, and therefore the angles SFD, SKP are equal.

Also the angle FSD

$$= \text{complement of } DSP$$

$$= SPK;$$

$\therefore$ the triangles FSD, SPK are similar, and

$$SD : SF :: SP : PK$$

$$:: SA : AX.$$

Cor. (1). If the tangent at the other end P' of the focal chord meet the directrix in D',

$$SD' : SF :: SA : AX;$$

$$\therefore SD = SD'.$$

Cor. (2). If DE be the perpendicular from D upon SP, the triangles SDE, SFX are similar, and

$$SE : SX :: SD : SF$$

$$:: SA : AX$$

$$:: SB : SX\,;$$

$\therefore$ SE is equal to SR, the semi-latus rectum.

15. Prop. XIII. *The tangents drawn from any point to a conic subtend equal angles at the focus.*

Let the tangents FTP, $F'TP'$ at P and P' meet the directrix in F and F' and the latus rectum in D and D'.

Join ST and produce it to meet the directrix in K;

then $$KF : SD :: KT : ST$$

$$:: KF' : SD'.$$

Hence $$KF : KF' :: SD : SD'$$

$$:: SF : SF' \text{ by Prop. XII.}$$

$\therefore$ the angles TSF, TSF' are equal.

But the angles FSP', $F'SP$ are equal, for each is the complement of FSF';

$\therefore$ the angles TSP, TSP' are equal.

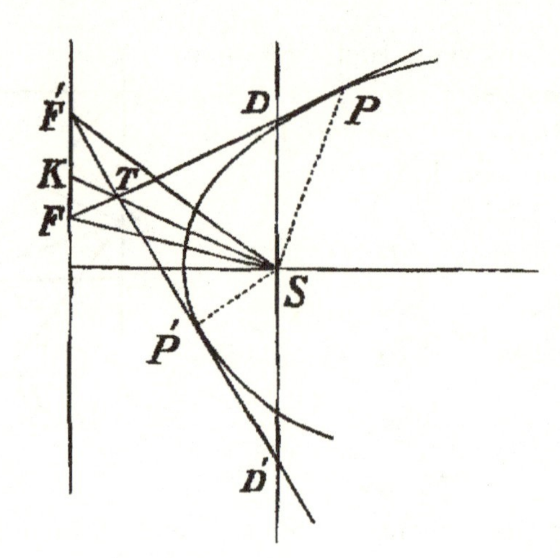

Cor. Hence it follows that if perpendiculars TM, TM' be let fall upon SP and SP', they are equal in length.

For the two triangles TSM, TSM' have the angles TMS, TSM respectively equal to the angles $TM'S$, TSM', and the side TS common; and therefore the other sides are equal, and

$$TM = TM'.$$

16. Prop. XIV. *If from any point T in the tangent at a point P of a conic, TM be drawn, perpendicular to the focal distance SP, and TN perpendicular to the directrix,*

$$SM : TN :: SA : AX.$$

For, if PK be perpendicular to the directrix and SF be joined,

$$SM : SP :: TF : FP$$
$$:: TN : PK;$$
$$\therefore SM : TN :: SP : PK$$
$$:: SA : AX.$$

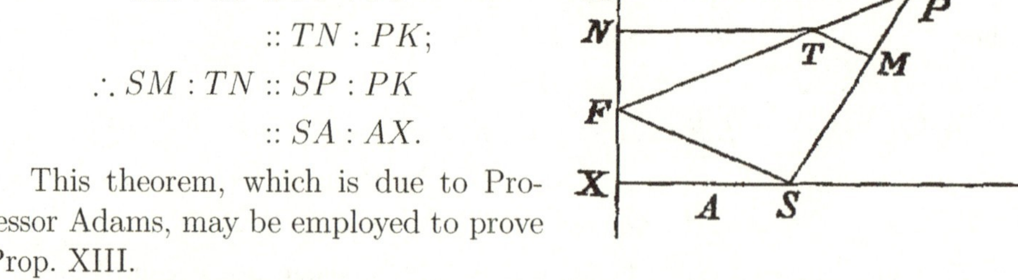

This theorem, which is due to Professor Adams, may be employed to prove Prop. XIII.

For if, in the figure of Art. (15), TM, TM' be the perpendiculars from T on SP and SP', and if TN be the perpendicular on the directrix, SM and SM' have each the same ratio to TN, and are therefore equal to one another.

Hence the triangles TSM, TSM' are equal in all respects, and the angle PSP' is bisected by ST.

17. PROP. XV. *To draw tangents from any point to a conic.*

Let T be the point, and let a circle be described about S as centre, the radius of which bears to TN the ratio of $SA : AX$; then, if tangents TM, TM' be drawn to the circle, the straight lines SM, SM', produced if necessary, will intersect the conic in the points of contact of the tangents from T.

18. PROP. XVI. *If PG, the normal at P, meet the axis of the conic in G,*

$$SG : SP :: SA : AX.$$

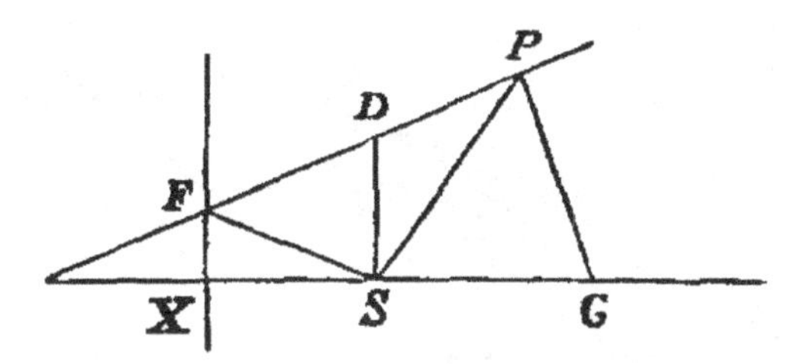

Let the tangent at P meet the directrix in F, and the latus rectum produced in D.

Then the angle SPG = the complement of $SPF = PFS$, and PSG = the complement of $FSX = FSD$;

$\therefore$ the triangles SFD, SPG are similar, and

$$SG : SP :: SD : SF :: SA : AX, \text{ by Prop. XII.}$$

19. PROP. XVII. *If from G, the point in which the normal at P meets the axis, GL be drawn perpendicular to SP, the length PL is equal to the semi-latus rectum.*

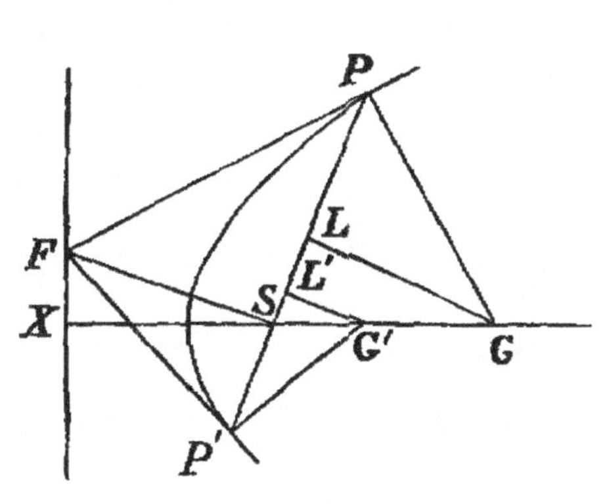

Let the tangent at P meet the directrix in F, and join SF.

Then PLG, PSF are similar triangles;

$$\therefore PL : LG :: SF : SP.$$

Also SLG and SFX are similar triangles;

$$\therefore LG : SX :: SG : SF.$$

Hence $$PL : SX :: SG : SP$$
$$:: SA : AX, \text{ Art. (18)},$$
but $$SR : SX :: SA : AX, \text{ Art. (2)};$$
$$\therefore PL = SR.$$

20. Prop. XVIII. *If from any point F in the directrix tangents be drawn, and also any straight line FPP' cutting the curve in P and P', the chord PP' is divided harmonically at F and its point of intersection with the chord of contact.*

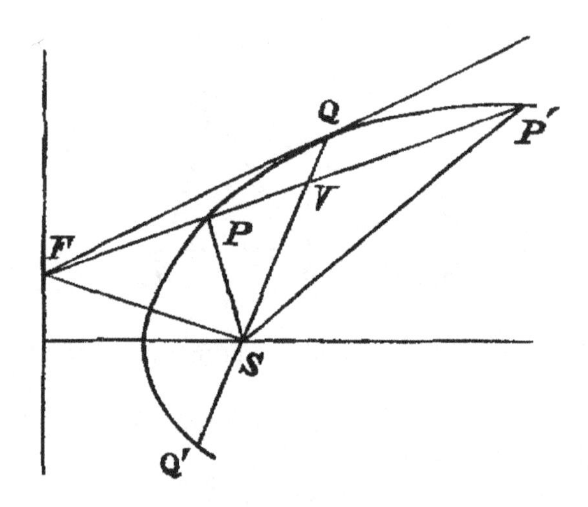

For, if QSQ' be the chord of contact, it bisects the angle PSP', (Prop. XI.), and $\therefore$, if V be the point of intersection of SQ and PP',

$$FP' : FP :: SP' : SP$$
$$:: P'V : PV$$
$$:: FP' - FV : FV - FP.$$

Hence FV is the harmonic mean between FP and FP'.

The theorems of this article and of Art. 9 are particular cases of more general theorems, which will appear hereafter.

21. Prop. XIX. *If a tangent be drawn parallel to a chord of a conic, the portion of this tangent which is intercepted by the tangents at the ends of the chord is bisected at the point of contact.*

Let PP' be the chord, TP, TP' the tangents, and EQE' the tangent parallel to PP'.

From the focus S draw SP, SP' and SQ, and draw TM, TM' perpendicular respectively to SP, SP'.

Also draw from E perpendiculars EN, EL, upon SP, SQ, and from E' perpendiculars $E'N'$, $E'L'$ upon SP' and SQ.

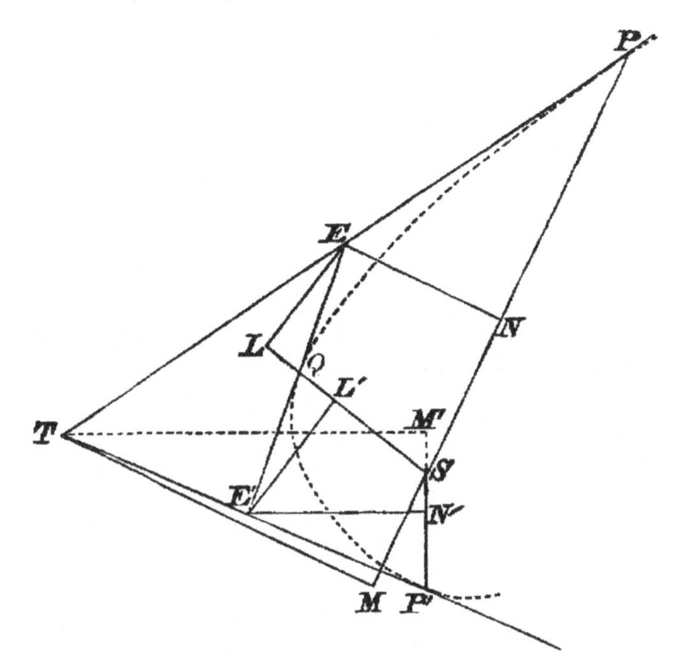

Then, since EE' is parallel to PP'

$$TP : EP :: TP' : E'P',$$

but
$$TP : EP :: TM : EN,$$

and
$$TP' : E'P' :: TM' : E'N';$$

$$\therefore TM : EN :: TM' : E'N';$$

but $TM = TM'$, Cor. Prop. XIII.;

$$\therefore EN = E'N'.$$

Again, by the same corollary,

$$EN = EL \text{ and } E'N' = E'L';$$

$$\therefore EL = E'L',$$

and, the triangles ELQ, $E'L'Q$ being similar,

$$EQ = E'Q.$$

COR. If TQ be produced to meet PP' in V,

$$PV : EQ :: TV : TQ,$$

and

$$P'V : E'Q :: TV : TQ;$$

$$\therefore PV = P'V,$$

that is, PP' is bisected in V.

Hence, if tangents be drawn at the ends of any chord of a conic, the point of intersection of these tangents, the middle point of the chord, and the point of contact of the tangent parallel to the chord, all lie in one straight line.

EXAMPLES.

1. Describe the relative positions of the focus and directrix, first, when the conic is a circle, and secondly, when it consists of two straight lines.

2. Having given two points of a conic, the directrix, and the eccentricity, determine the conic.

3. Having given a focus, the corresponding directrix, and a tangent, construct the conic.

4. If a circle passes through a fixed point and cuts a given straight line at a constant angle the locus of its centre is a conic.

5. If PG, pg, the normals at the ends of a focal chord, intersect in O, the straight line through O parallel to Pp bisects Gg.

6. Find the locus of the foci of all the conics of given eccentricity which pass through a fixed point P, and have the normal PG given in magnitude and position.

7. Having given a point P of a conic, the tangent at P, and the directrix, find the locus of the focus.

8. If PSQ be a focal chord, and X the foot of the directrix, XP and XQ are equally inclined to the axis.

9. If PK be the perpendicular from a point P of a conic on the directrix, and SK meet the tangent at the vertex in E, the angles SPE, KPE are equal.

10. If the tangent at P meet the directrix in F and the axis in T, the angles KSF, FTS are equal.

11. PSP' is a focal chord, PN, $P'N'$ are the ordinates, and PK, $P'K'$ perpendiculars on the directrix; if KN, $K'N'$ meet in L, the triangle LNN' is isosceles.

12. The focal distance of a point on a conic is equal to the length of the ordinate produced to meet the tangent at the end of the latus rectum.

13. The normal at any point bears to the semi-latus rectum the ratio of the focal distance of the point to the distance of the focus from the tangent.

14. The chord of a conic is given in length; prove that, if this length exceed the latus rectum, the distance from the directrix of the middle point of the chord is least when the chord passes through the focus.

15. The portion of any tangent to a conic, intercepted between two fixed tangents, subtends a constant angle at the focus.

16. Given two points of a conic, and the directrix, find the locus of the focus.

17. From any fixed point in the axis a line is drawn perpendicular to the tangent at P and meeting SP in R; the locus of R is a circle.

18. If the tangent at the end of the latus rectum meet the tangent at the vertex in T, $AT = AS$.

19. TP, TQ are the tangents at the points P, Q of a conic, and PQ meets the directrix in R; prove that RST is a right angle.

20. SR being the semi-latus rectum, if RA meet the directrix in E, and SE meet the tangent at the vertex in T,

$$AT = AS.$$

21. If from any point T, in the tangent at P, TM be drawn perpendicular to SP, and TN perpendicular to the transverse axis, meeting the curve in R, $SM = SR$.

22. If the chords PQ, $P'Q$ meet the directrix in F and F', the angle FSF' is half PSP'.

23. If PN be the ordinate, PG the normal, and GL the perpendicular from G upon SP,

$$GL : PN :: SA : AX.$$

24. If normals be drawn at the ends of a focal chord, a line through their intersection parallel to the axis will bisect the chord.

25. If a conic of given eccentricity is drawn touching the straight line FD joining two fixed points F and D, and if the directrix always passes through F, and the corresponding latus rectum always passes through D, find the locus of the focus.

26. If ST, making a constant angle with SP meet in T the tangent at P, prove that the locus of T is a conic having the same focus and directrix.

27. If E be the foot of the perpendicular let fall upon PSP' from the point of intersection of the normals at P and P',

$$PE = SP' \text{ and } P'E = SP.$$

28. If a circle be described on the latus rectum as diameter, and if the common tangent to the conic and circle touch the conic in P and the circle in Q, the angle PSQ is bisected by the latus rectum. (Refer to Cor. 2. Art. 14.)

29. Given two points, the focus, and the eccentricity, determine the position of the axis.

30. If a chord PQ subtend a constant angle at the focus, the locus of the intersection of the tangents at P and Q is a conic with the same focus and directrix.

31. The tangent at a point P of a conic intersects the tangent at the fixed point P' in Q, and from S a straight line is drawn perpendicular to SQ and meeting in R the tangent at P; prove that the locus of R is a straight line.

32. The circle is drawn with its centre at S, and touching the conic at the vertex A; if radii Sp, Sp' of the circle meet the conic in P, P', prove that PP', pp' intersect on the tangent at A.

33. Pp is any chord of a conic, PG, pg the normals, G, g being on the axis; GK, gk are perpendiculars on Pp; prove that $PK = pk$.

CHAPTER II.

THE PARABOLA.

DEF. *A parabola is the curve traced out by a point which moves in such a manner that its distance from a given point is always equal to its distance from a given straight line.*

Tracing the Curve.

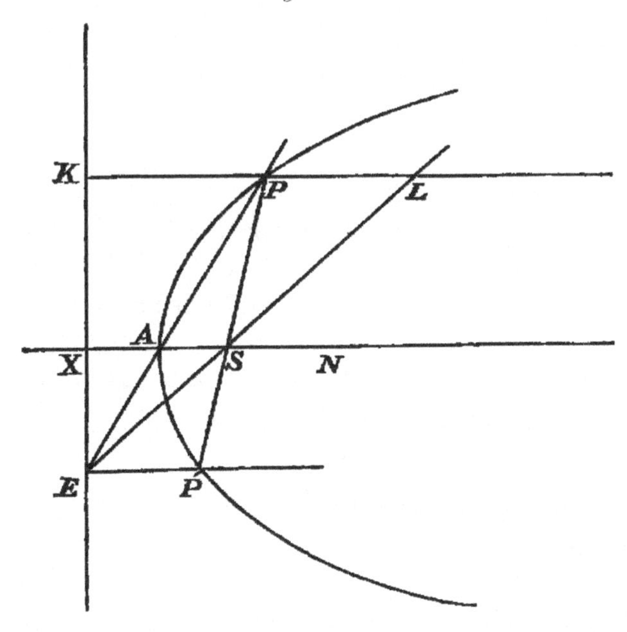

22. Let S be the focus, EX the directrix, and SX the perpendicular on EX. Then, bisecting SX in A, the point A is the vertex; and if, from any

point E in the directrix, EAP, ESL be drawn, and from S the straight line SP meeting EA produced in P, and making the angle PSL equal to LSN, we obtain, as in Art. (1), a point P in the curve.

For $$PL : PK :: SA : AX,$$

and $$\therefore PL = PK.$$

But $$SP = PL, \text{ and } \therefore SP = PK.$$

Again, drawing EP' parallel to the axis and meeting in P' the line PS produced, we obtain the other extremity of the focal chord PSP'.

For the angle $$ESP' = PSL = PLS$$ $$= SEP',$$

and $$\therefore SP' = P'E,$$

and P' is a point in the parabola.

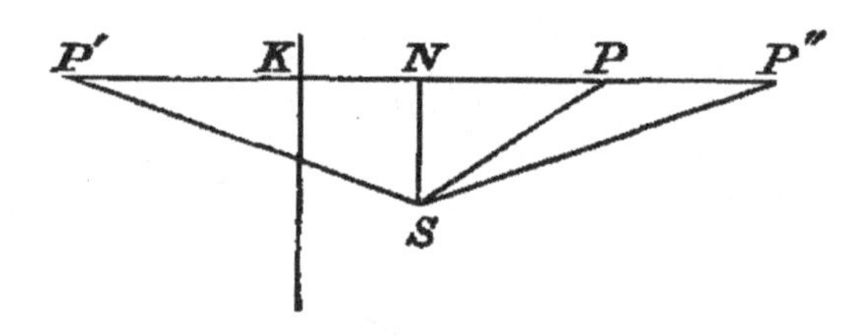

The curve lies wholly on the same side of the directrix; for, if P' be a point on the other side, and SN be perpendicular to $P'K$, SP' is greater than $P'N$, and therefore is greater than $P'K$.

Again, a straight line parallel to the axis meets the curve in one point only.

For, if possible, let P'' be another point of the curve in KP produced.

Then $$SP = PK \text{ and } SP'' = P''K$$ $$\therefore PP'' = SP'' - SP,$$

or $$PP'' + SP = SP'',$$

which is impossible.

23. Prop. I. *The distance from the focus of a point inside a parabola is less, and of a point outside is greater than its distance from the directrix.*

If Q be the point inside, let fall the perpendicular QPK on the directrix, meeting the curve in P.

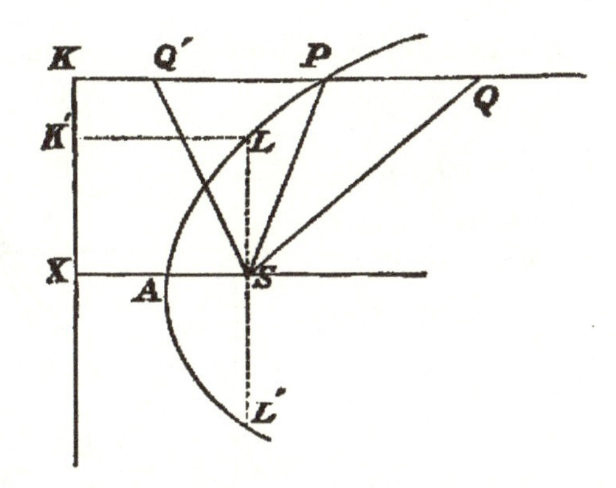

Then $SP + PQ > SQ$, but

$$SP + PQ = PK + PQ = QK,$$

$$\therefore SQ < QK.$$

If Q' be outside, and between P and K,

$$SQ' + PQ' > SP,$$

$$\therefore SQ' > Q'K.$$

If Q' lie in PK produced,

$$SQ' + SP > PQ',$$

and

$$\therefore SQ' > KQ'.$$

24. PROP. II. *The latus rectum* $= 4 \,.\, AS$.

For if, Fig. Art. 23, LSL' be the latus rectum, drawing LK' at right angles to the directrix, we have

$$LS = LK' = SX = 2AS,$$

$$\therefore LSL' = 4 \,.\, AS.$$

25. *Mechanical construction of the Parabola.*

Take a rigid bar EKL, of which the portions EK, KL are at right angles to each other, and fasten a string to the end L, the length of which is LK. Then if the other end of the string be fastened to S, and the bar be made to slide along a fixed straight edge, EKX, a pencil at P, keeping the string stretched against

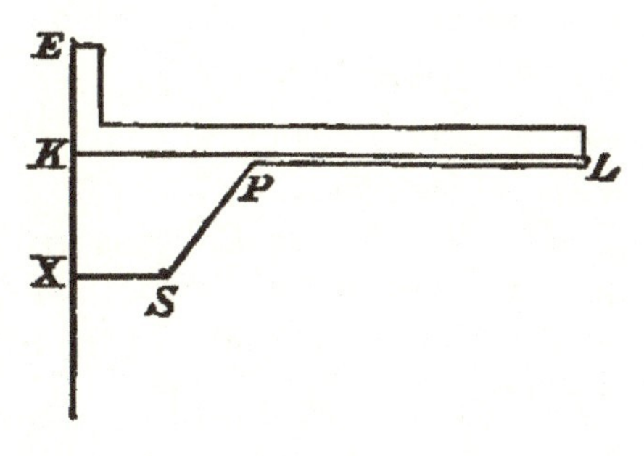

the bar, will trace out a portion of a parabola, of which S is the focus, and EX the directrix.

26. Prop. III. *If PK is the perpendicular upon the directrix from a point P of a parabola, and if PA meet the directrix in E, the angle KSE is a right angle.*

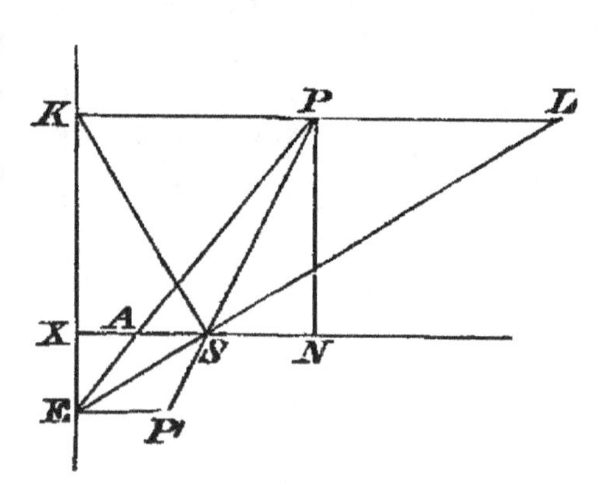

Join ES, and let KP and ES produced meet at L.

Since $SA = AX$, it follows that $PL = PK = SP$;

$\therefore$ P is the centre of the circle through K, S, and L, and the angle KSL is a right angle.

Therefore KSE is a right angle.

27. Prop. IV. *If PN is the ordinate of a point P of a parabola,*

$$PN^2 = 4AS \,.\, AN.$$

Taking the figure above,

$$PN : EX :: AN : AX$$

$$\therefore PN^2 : EX \,.\, KX :: 4AS \,.\, AN : 4AS^2.$$

But, since KSE is a right angle,

$$EX \,.\, KX = SX^2 = 4AS^2,$$

$$\therefore PN^2 = 4AS \,.\, AN.$$

Cor. If AN increases, and becomes infinitely large, PN increases and becomes infinitely large, and therefore the two portions of the curve, above and below the axis, proceed to infinity.

28. Prop. V. *If from the ends of a focal chord perpendiculars be let fall upon the directrix, the intercepted portion of the directrix subtends a right angle at the focus.*

For, if PA meet the directrix in E, and if the straight line through E perpendicular to the directrix meet PS in P', it is shewn, in Art. 22, that P' is the other extremity of the focal chord PS; and, as in Art. 26, KSE is a right angle.

29. Prop. VI. *The tangent at any point P bisects the angle between the focal distance SP and the perpendicular PK on the directrix.*

Let F be the point in which the tangent meets the directrix, and join SF.

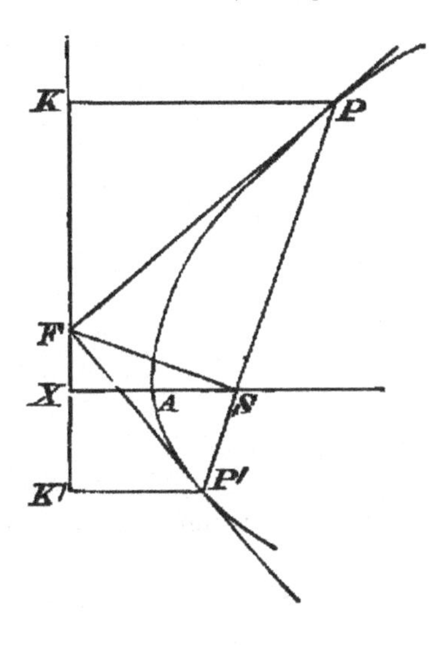

We have shewn, (Art. 10) that FSP is a right angle, and, since $SP = PK$, and PF is common to the right-angled triangles SPF, KPF, it follows that these triangles are equal in all respects, and therefore the angle

$$SPF = FPK.$$

In other words, *the tangent at any point is equally inclined to the focal distance and the axis.*

Cor. It has been shewn, in Art. (12), that the tangents at the ends of a focal chord intersect in the directrix, and therefore, if PS produced meet the curve in P', FP' is the tangent at P', and bisects the angle between SP' and the perpendicular from P' on the directrix.

30. Prop. VII. *The tangents at the ends of a focal chord intersect at right angles in the directrix.*

Let PSP' be the chord, and PF, $P'F$ the tangents meeting the directrix in F.

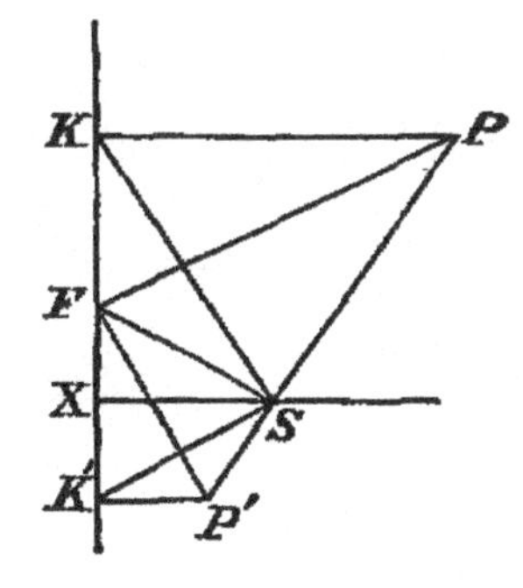

Let fall the perpendiculars PK, $P'K'$, and join SK, SK'.

The angle $P'SK' = \frac{1}{2}P'SX$
$= \frac{1}{2}SPK = SPF$,

$\therefore SK'$ is parallel to PF,

and, similarly, SK is parallel to $P'F$.

But (Art. 28) KSK' is a right angle;

$$\therefore PFP' \text{ is a right angle.}$$

31. Prop. VIII. *If the tangent at any point P of a parabola meet the axis in T, and PN be the ordinate of P, then*

$$AT = AN.$$

Draw PK perpendicular to the directrix.

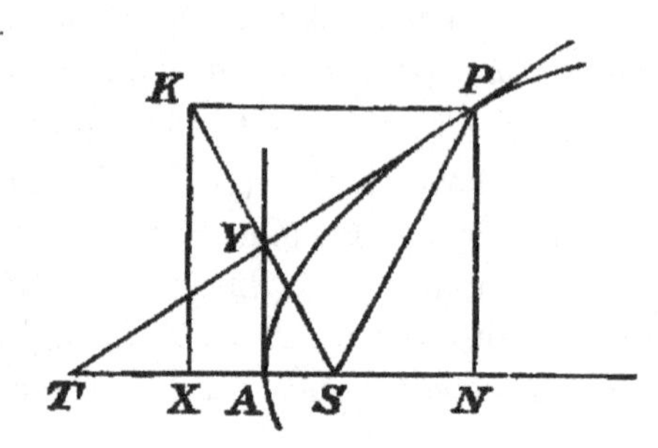

The angle $SPT = TPK$

$$= PTS,$$

$$\therefore ST = SP$$

$$= PK$$

$$= NX.$$

But $$ST = SA + AT,$$

and $$NX = AN + AX;$$

$$\therefore \text{ since } SA = AX,$$

$$AT = AN.$$

DEF. *The line NT is called the sub-tangent.*

The sub-tangent is therefore twice the abscissa of the point of contact.

32. PROP. IX. *The foot of the perpendicular from the focus on the tangent at any point P of a parabola lies on the tangent at the vertex, and the perpendicular is a mean proportional between SP and SA.*

Taking the figure of the previous article, join SK meeting PT in Y.

Then $SP = PK$, and PY is common to the two triangles SPY, KPY;

$$\text{also the angle } SPY = YPK;$$

$$\therefore \text{ the angle } SYP = PYK,$$

and SY is perpendicular to PT.

Also $SY = KY$, and $SA = AX$, $\therefore AY$ is parallel to KX.

Hence, AY is at right angles to AS, and is therefore the tangent at the vertex.

Again, the angle $SPY = STY = SYA$, and the triangles SPY, SYA are therefore similar;

$$\therefore SP : SY :: SY : SA,$$

$$\text{or } SY^2 = SP \,.\, SA.$$

33. PROP. X. *In the parabola the subnormal is constant and equal to the semi-latus rectum.*

DEF. *The distance between the foot of the ordinate of P and the point in which the normal at P meets the axis is called the subnormal.*

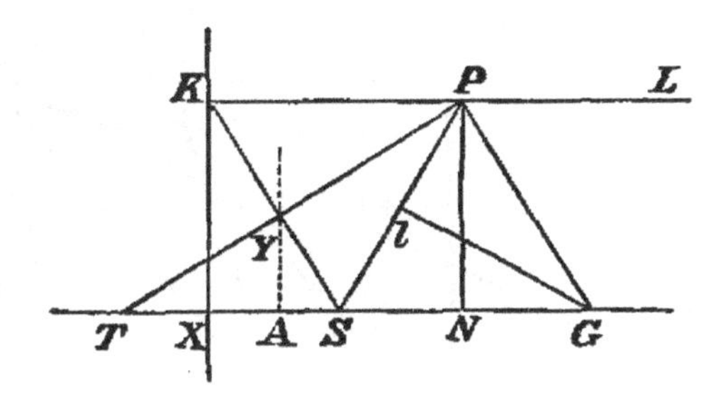

In the figure PG is the normal and PT the tangent.

It has been shewn that the angle SPK is bisected by PT, and hence it follows that SPL is bisected by PG,

and that the angle $\quad SPG = GPL = PGS$;

hence

$$\begin{aligned} SG &= SP = ST \\ &= SA + AT = SA + AN \\ &= 2AS + SN; \end{aligned}$$

$\therefore$ the subnormal $NG = 2AS$.

34. Cor. If Gl be drawn perpendicular to SP,
the angle GPl = the complement of SPT,
= the complement of STP,
$= PGN$,

and the two right-angled triangles GPN, GPl have their angles equal and the side GP common; hence the triangles are equal, and

$$Pl = NG = 2AS$$
$$= \text{the semi-latus rectum.}$$

It has been already shewn, (Art. 19), that this property is a general property of all conics.

35. Prop. XI. *To draw tangents to a parabola from an external point.*

For this purpose we may employ the general construction given in Art. (17), or, for the special case of the parabola, the following construction.

Let Q be the external point, join SQ, and upon SQ as diameter describe a circle intersecting the tangent at the vertex in Y and Y'. Join YQ, $Y'Q$; these are tangents to the parabola.

Draw SP, so as to make the angle YSP equal to YSA, and to meet YQ in P, and let fall the perpendicular PN upon the axis.

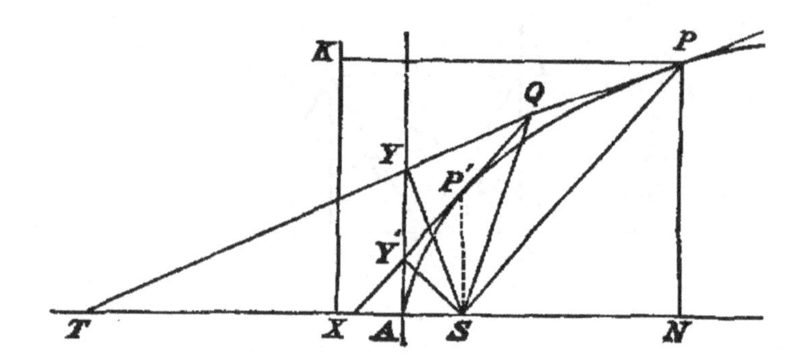

Then, SYQ is a right angle, since it is the angle in a semicircle, and, T being the point in which QY produced meets the axis, the two triangles SYP, SYT are equal in all respects;

$$\therefore SP = ST, \text{ and } YT = YP.$$

But AY is parallel to PN;

$$\therefore AT = AN.$$

Hence
$$\begin{aligned} SP = ST &= SA + AT \\ &= AX + AN \\ &= NX, \end{aligned}$$

and P is a point in the parabola.

Moreover, if PK be perpendicular to the directrix, the angle $SPY = STP = YPK$, and PY is the tangent at P. (Art. 29.)

Similarly, by making the angle $Y'SP'$ equal to ASY' we obtain the point of contact of the other tangent QY'.

36. Prop. XII. *If from a point Q tangents QP, QP' be drawn to a parabola, the two triangles SPQ, SQP', are similar, and SQ is a mean proportional between SP and SP'.*

Produce PQ to meet the axis in T, and draw SY, SY' perpendicularly on the tangents. Then Y and Y' are points in the tangent at A.

The angle
$$\begin{aligned} SPQ &= STY \\ &= SYA \\ &= SQP', \end{aligned}$$

since S, Y', Y, Q are points on a circle, and SYA, SQP' are in the same segment.

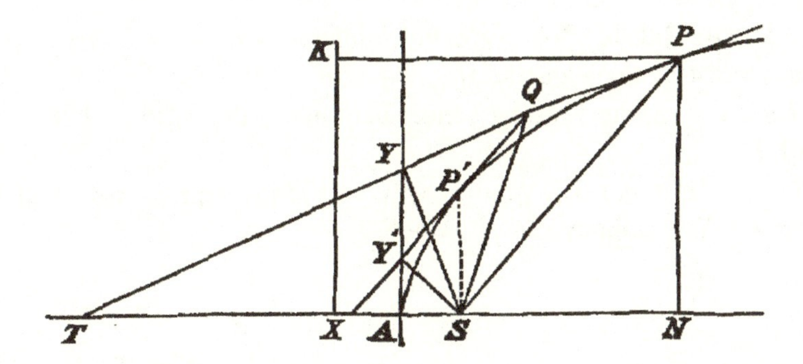

Also, by the theorem of Art. (15), the angle

$$PSQ = QSP';$$

therefore the triangles PSQ, QSP' are similar, and

$$SP : SQ :: SQ : SP'.$$

37. From the preceding theorem the following, which is often useful, immediately follows.

If from any points in a given tangent of a parabola, tangents be drawn to the curve, the angles which these tangents make with the focal distances of the points from which they are drawn are all equal.

For each of them by the theorem, is equal to the angle between the given tangent and the focal distance of the point of contact.

Hence it follows that the locus of the intersection of a tangent to a parabola with a straight line drawn through the focus meeting it at a constant angle is a straight line.

For if QP be the moveable tangent, the angle $SQP = SP'Q$, and therefore, if SQP is constant, $SP'Q$ is a given angle. The point P' is therefore fixed, and the locus of Q is the tangent $P'Q$.

38. Since the two triangles PSQ, QSP' are similar, we have

$$PQ : P'Q :: SP : SQ$$

and

$$PQ : P'Q :: SQ : SP',$$

$$\therefore PQ^2 : P'Q^2 :: SP : SP';$$

that is, the squares of the tangents from any point are proportional to the focal distances of the points of contact.

This will be found to be a particular case of a subsequent Theorem, given in Art. 51.

39. Prop. XIII. *The external angle between two tangents is half the angle subtended at the focus by the chord of contact.*

Let the tangents at P and P' intersect each other in Q and the axis ASN in T and T'.

Join SP, SP'; then the angles SPT, STP are equal, and $\therefore STP$ is half the angle PSN; similarly $ST'P'$ is half $P'SN$.

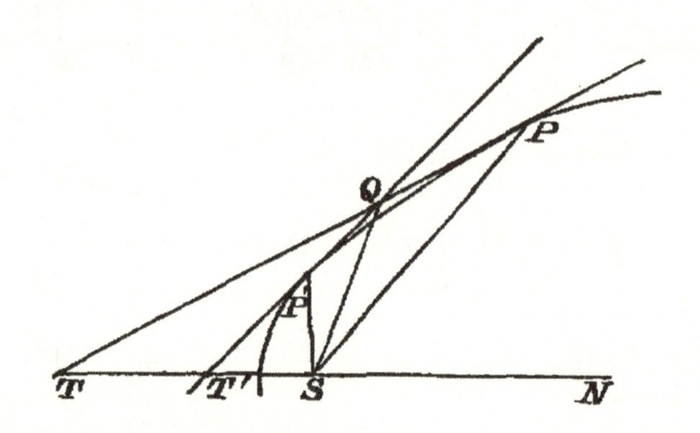

But TQT' is equal to the difference between STP and $ST'P'$, and is therefore equal to half the difference between PSN and $P'SN$, that is to half the angle PSP'.

Hence, joining SQ, TQT' is equal to each of the angles PSQ, $P'SQ$.

40. Prop. XIV. *The tangents drawn to a parabola from any point make the same angles, respectively, with the axis and the focal distance of the point.*

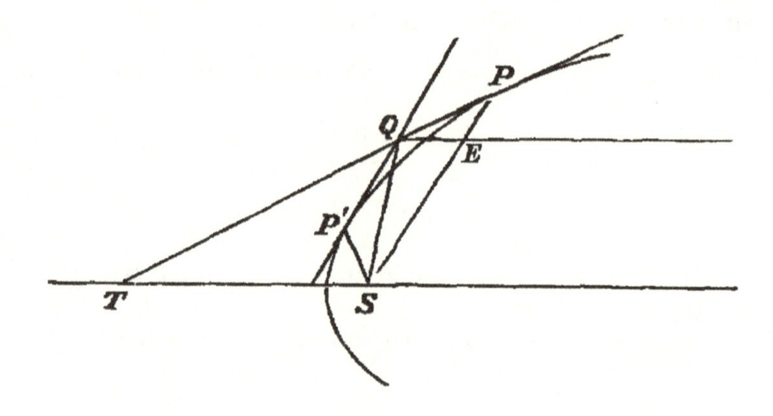

Let QP, QP' be the tangents; join SP, and draw QE parallel to the axis, and meeting SP in E.

Then, if PQ meet the axis in T, the angle

$$\begin{aligned} EQP &= STP = SPQ \\ &= SQP'. \quad \text{(Art. 37.)} \end{aligned}$$

i.e. QP and QP' respectively make the same angles with the axis and with QS.

41. Conceive a parabola to be drawn passing through Q, having S for its focus, SN for its axis, and its vertex on the same side of S as the vertex A of the given parabola. Then the normal at Q to this new parabola bisects the angle SQE; therefore the angles which QP and QP' make with the normal at Q are equal.

Hence the theorem,

If from any point in a parabola, tangents be drawn to a confocal and co-axial parabola, the normal at the point will bisect the angle between the tangents.

If we produce SP to any point p, and take St equal to Sp, pt will be the tangent at p to the confocal and co-axial parabola passing through p.

Hence the theorem,

If parallel tangents be drawn to a series of confocal and co-axial parabolas, the points of contact will lie in a straight line passing through the focus.

In these enunciations the words co-axial and confocal are intended to imply, not merely the coincidence of the axes, but also that the vertices of the two parabolas are on the same side of their common focus.

The reason for this will appear when we shall have discussed the analogous property of the ellipse.

42. If two confocal parabolas have their axes in the same straight line, and their vertices on opposite sides of the focus, they intersect at right angles.

For the angle $\qquad TPS = \frac{1}{2}PST'$,

and $\qquad T'PS = \frac{1}{2}PST$,

$$\therefore TPT' = \tfrac{1}{2}(PST + PST') = \text{a right angle.}$$

It will be noticed that, in this case, the common chord PQ is equidistant from the directrices.

For the distance of P from each directrix is equal to SP.

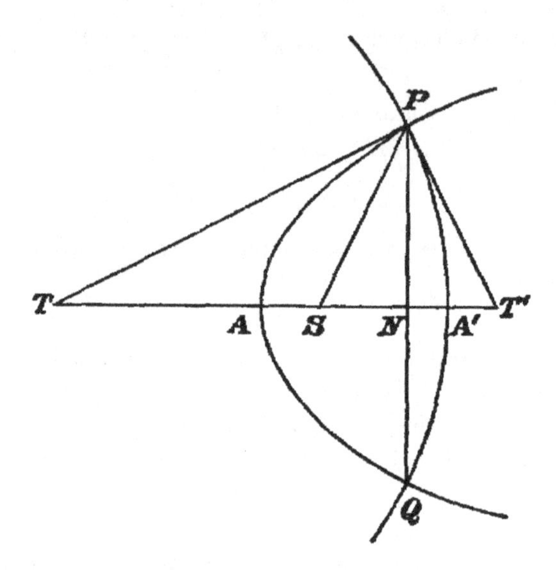

43. Prop. XV. *The circle passing through the points of intersection of three tangents passes also through the focus.*

Let Q, P, Q' be the three points of contact, and F, T, F' the intersections of the tangents.

In Art. (36) it has been shewn that, if FP, FQ be tangents, the angle

$$SQF = SFP.$$

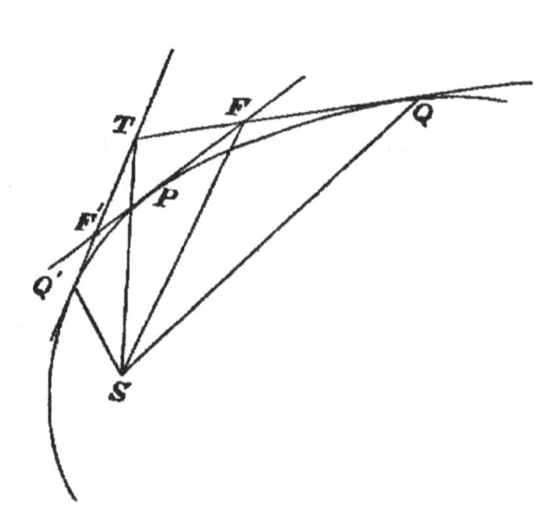

Similarly TQ, TQ' being tangents, the angle

$$SQT = STQ',$$

hence the angle $\qquad SFF'$ or $SFP = SQT,$

$$= STF',$$

and a circle can be drawn through S, F, T, and F'.

44. DEF. *A straight line drawn parallel to the axis through any point of a parabola is called a diameter.*

PROP. XVI. *If from any point T tangents TQ, TQ' be drawn to a parabola, the point T is equidistant from the diameters passing through Q and Q', and the diameter drawn through the point T bisects the chord of contact.*

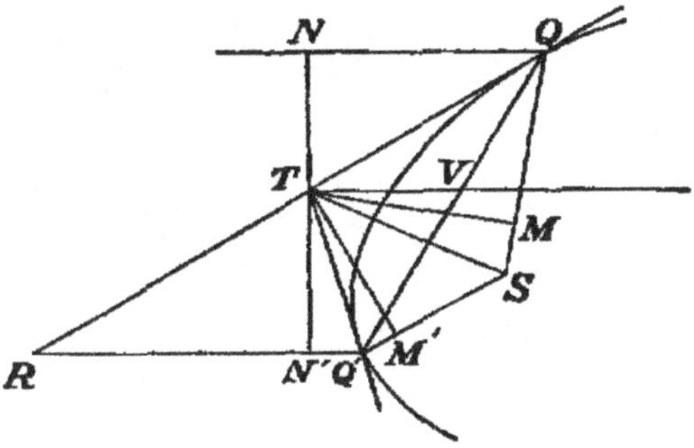

Join SQ, SQ', and draw TM, TM' perpendicular respectively to SQ and SQ'.

Also draw NTN' perpendicular to the diameters through Q and Q', and meeting those diameters in N and N'.

Then, since TS bisects the angle QSQ',

$$TM = TM';$$

and, since TQ bisects the angle SQN,

$$TN = TM.$$

Similarly $\qquad TN' = TM',$

$$\therefore TN = TN'.$$

Again, join QQ', and draw the diameter TV meeting QQ' in V; also let QT produced meet $Q'N'$ in R;

then $\qquad QV : VQ' :: QT : TR$

$$:: TN : TN',$$

since the triangles QTN, RTN' are similar;

$$\therefore QV = VQ'.$$

Hence *the diameter through the middle point of a chord passes, when produced, through the point of intersection of the tangents at the ends of the chord.*

It should be noticed that any straight line drawn through T and terminated by QN and $Q'N'$ is bisected at T.

45. PROP. XVII. *Any diameter bisects all chords parallel to the tangent at its extremity, and passes through the point of intersection of the tangents at the ends of any of these chords.*

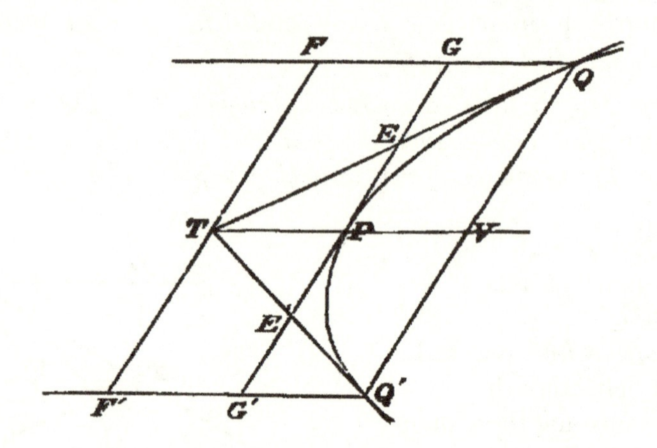

Let QQ' be a chord parallel to the tangent at P, and through the point of intersection T of the tangents at Q and Q' draw FTF' parallel to QQ' and terminated at F and F' by the diameters through Q and Q'.

Let the tangent at P meet TQ, TQ' in E and E', and QF, $Q'F'$ in G and G'.

Then
$$EG : TF :: EQ : TQ$$
$$:: E'Q' : TQ'$$
$$:: E'G' : TF'.$$

But $TF = TF'$, since (Art. 44) T is equidistant from QG and $Q'G'$,
$$\therefore EG = E'G'.$$

Also, $EP = EG$, since E is equidistant from QG and PV, the diameter at P.

$$\therefore EP = E'P \text{ and } GP = PG',$$

and
$$\therefore QV = VQ'.$$

Again, since T, P, V are each equidistant from the parallel straight lines QF, $Q'F'$, it follows that TPV is a straight line, or that the diameter VP passes through T.

We have shewn that GE, EP, PE', $E'G'$ are all equal, and we hence infer that

$$EE' = \tfrac{1}{2}GG' = \tfrac{1}{2}QQ',$$

and consequently that $TP = \frac{1}{2}TV$, or that $TP = PV$.

Hence *it appears, that the diameter through the point of intersection of a pair of tangents passes through the point of contact of the tangent parallel to the chord of contact, and also through the middle point of the chord of contact; and that the portion of the diameter between the point of intersection of the tangents and the middle point of the chord of contact is bisected at the point of contact of the parallel tangent.*

We may observe that in proving that EE' is bisected at P, we have demonstrated a theorem already shewn (Art. 21) to be true for all conics.

46. When the point T is on the directrix, QTQ' is a right angle.

If then Qq is the chord which is normal at Q, it is parallel to the tangent TQ', and is therefore bisected by the diameter $Q'U$ through Q'.

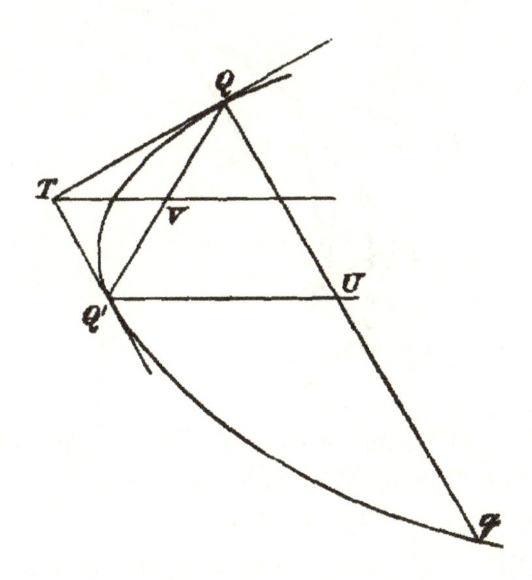

Since QU is bisected by TV, it follows that

$$Qq = 4TQ',$$

i.e. *the length of a normal chord is four times the portion of the parallel tangent between the directrix and the point of contact.*

47. DEF. *The line QV, parallel to the tangent at P, and terminated by the diameter PV, is called an ordinate of that diameter, and QQ' is the double ordinate. The point P, the end of the diameter, is called the vertex of the diameter, and the distance PV is called the abscissa of the point Q.*

We have seen that tangents at the ends of any chord intersect in the diameter which bisects the chord, and that the distance of this point from the vertex is equal to the distance of the vertex from the middle point of the chord.

DEF. *The chord through the focus parallel to the tangent at any point is called the parameter of the diameter passing through the point.*

PROP. XVIII. *The parameter of any diameter is four times the focal distance of the vertex of that diameter.*

Let P be the vertex, and QSQ' the parameter, T the point of intersection of the tangents at Q and Q', and FPF' the tangent at P.

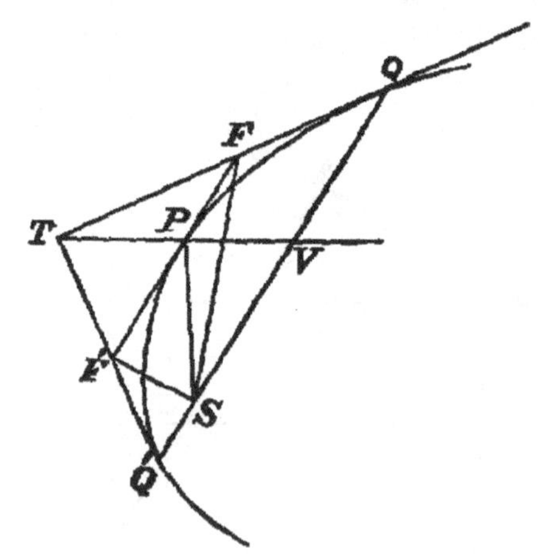

Then, since FS and $F'S$ bisect respectively the angles PSQ, PSQ', FSF' is a right angle, and, P being the middle point of FF', $SP = PF = PF'$.

Hence QQ', which is double FF', is four times SP.

48. PROP. XIX. *If QVQ' be a double ordinate of a diameter PV, QV is a mean proportional between PV and the parameter of P.*

Let FPF' be the tangent at P, and draw the parameter through S meeting PV in U.

The angle $SUT = FPU = SPF'$ (Art. 29), and, since the angles SFQ, SPF are equal (Art. 36), it follows that the angles SFT, SPF' are equal;

$\therefore SUT = SFT$, and U is a point in the circle passing through $SFTF'$.

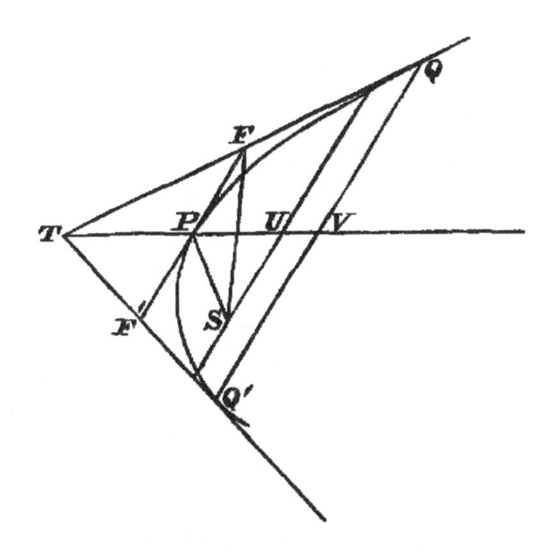

Hence, QV being twice PF,

$$QV^2 = 4PF^2 = 4PU\,.\,PT;$$

but $$PU = SP,$$

for the angle $$SUP = FPU = SPF' = PSU;$$

and $$PT = PV,$$

$$\therefore QV^2 = 4SP\,.\,PV.$$

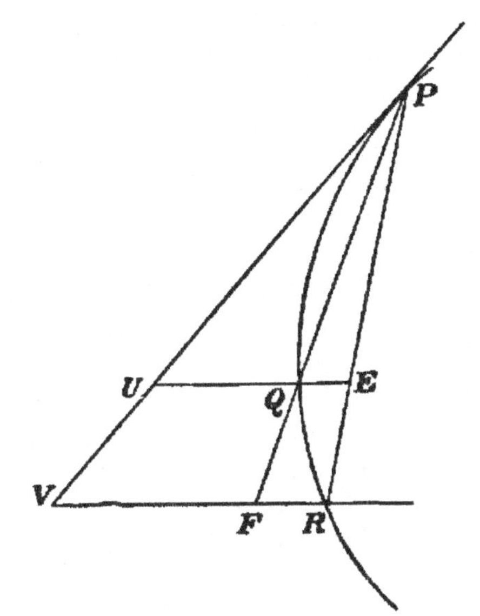

49. This relation may be presented in a different form, which is sometimes useful.

If from any point U in the tangent at P, UQ is drawn parallel to the axis, UP and UQ are respectively equal to the ordinate and abscissa of the point Q with regard to the diameter through P, and therefore

$$PU^2 = 4SP\,.\,UQ.$$

Therefore, if VR is drawn parallel to the axis from another point V of the tangent,

$$PU^2 : PV^2 :: UQ : VR.$$

Hence, since $UE : VR :: PU : PV,$

$$UE^2 : VR^2 :: UQ : VR :: UQ \, . \, VR : VR^2,$$

and $$UE^2 = UQ \, . \, VR.$$

Hence $$UE : UQ :: VR : UE :: PR : PE;$$

$$\therefore UQ : QE :: PE : ER.$$

In a similar manner it can be shewn that $VF^2 = UQ \, . \, VR$, and it follows that $VF = UE$, and therefore that EF is parallel to the tangent at P.

50. Prop. XX. *If QVQ' be a double ordinate of a diameter PV, and QD the perpendicular from Q upon PV, QD is a mean proportional between PV and the latus rectum.*

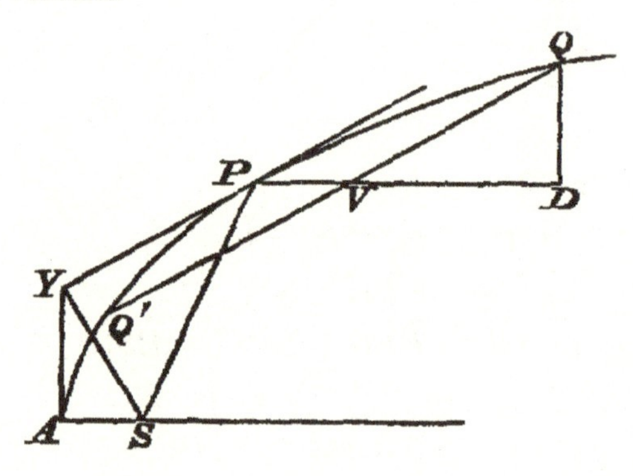

Let the tangent at P meet the tangent at the vertex in Y, and join SY.

The angle $QVD = SPY = SYA$, and therefore the triangles QVD, SAY are similar;

and $$QD^2 : QV^2 :: AS^2 : SY^2$$
$$:: AS^2 : AS \, . \, SP.$$
$$:: AS : SP$$
$$:: 4AS \, . \, PV : 4SP \, . \, PV,$$

but $$QV^2 = 4SP \, . \, PV;$$

$$\therefore QD^2 = 4AS \, . \, PV.$$

51. Prop. XXI. *If from any point, within or without a parabola, two straight lines be drawn in given directions and intersecting the curve, the ratio of the rectangles of the segments is independent of the position of the point.*

From any point O draw a straight line intersecting the parabola in Q and Q', and draw the diameter OE, meeting the curve in E.

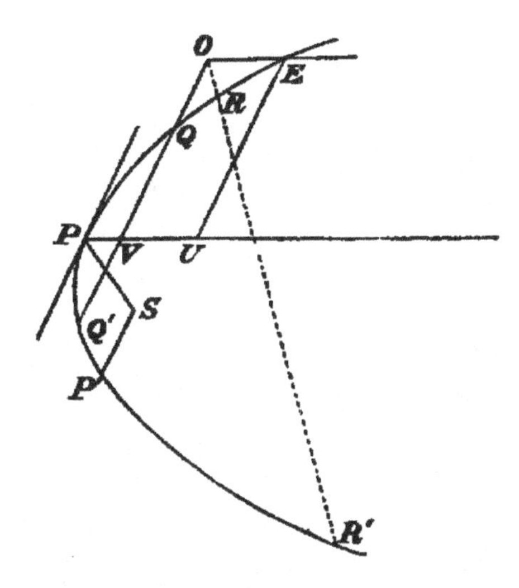

If PV be the diameter bisecting QQ', and EU the ordinate,

$$\begin{aligned} OQ\,.\,OQ' &= OV^2 - QV^2 \\ &= EU^2 - QV^2 = 4SP\,.\,PU - 4SP\,.\,PV \\ &= 4SP\,.\,OE. \end{aligned}$$

Similarly, if ORR' be any other intersecting line and P' the vertex of the diameter bisecting RR',

$$OR\,.\,OR' = 4SP'\,.\,OE.$$

$$\therefore OQ\,.\,OQ' : OR\,.\,OR' :: SP : SP',$$

that is, the ratio of the rectangles depends only on the positions of P and P', and, if the lines OQQ', ORR' are drawn parallel to given straight lines, these points P, P' are fixed.

It will be easily seen that the proof is the same if the point O be within the parabola.

If the lines OQQ', ORR' be moved parallel to themselves until they become the tangents at P and P', we shall then obtain, if these tangents intersect in T,

$$TP^2 : TP'^2 :: SP : SP';$$

a result previously obtained (Art. 38).

Again if QSQ', RSR' be the focal chords parallel to TP and TP', it follows that

$$TP^2 : TP'^2 :: QS\,.\,SQ' : RS\,.\,SR',$$

$\therefore$ (cor. Art. 8) $TP^2 : TP'^2 :: QQ' : RR'$.

52. Prop. XXII. *If from a point O, outside a parabola, a tangent OM, and a chord OAB be drawn, and if the diameter ME meet the chord in E,*

$$OE^2 = OA \,.\, OB.$$

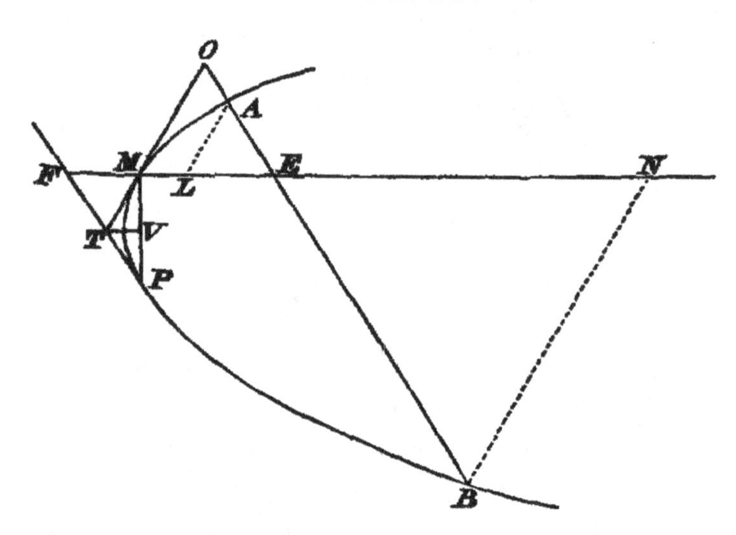

Let P be the point of contact of the tangent parallel to OAB, and let OM, ME meet this tangent in T and F.

Draw TV parallel to the axis and meeting PM in V;

then $$OA \,.\, OB : OM^2 :: TP^2 : TM^2 \text{ (Art. 51)},$$
$$:: TF^2 : TM^2,$$

since PM is bisected in V;

also $$TF : TM :: OE : OM;$$
$$\therefore OE^2 = OA \,.\, OB.$$

Cor. 1. If AL, BN be the ordinates, parallel to OM, of A and B, ML, ME, and MN are proportional to OA, OE and OB, and therefore

$$ME^2 = ML \,.\, MN.$$

This theorem may be also stated in the following form:

If a chord AB of a parabola intersect a diameter in the point E, the distance of the point E from the tangent at the end of the diameter is a mean proportional between the distances of the points A and B from the same tangent.

Cor. 2. Let KE be the ordinate through E parallel to OM.

Then, since $$ML : ME :: ME : MN,$$
$$AL^2 : KE^2 :: KE^2 : BN^2$$
$$\therefore AL : KE :: KE : BN,$$
so that KE is a mean proportional between AL and BN, the ordinates of A and B.

53. Prop. XXIII. *If a circle intersect a parabola in four points, the two straight lines constituting any one of the three pairs of the chords of intersection are equally inclined to the axis.*

Let Q, Q', R, R' be the four points of intersection;

then $$OQ \,.\, OQ' = OR \,.\, OR',$$
and therefore SP, SP' are equal, (Art. 51).

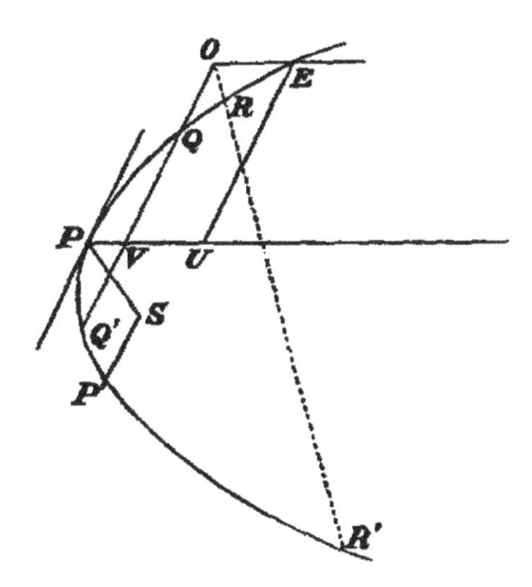

But, if SP, SP' be equal, the points P, P' are on opposite sides of, and are equidistant from the axis, and the tangents at P and P' are therefore equally inclined to the axis.

Hence the chords QQ', RR', which are parallel to these tangents, are equally inclined to the axis.

In the same manner it may be shewn that QR, $Q'R'$ are equally inclined to the axis, as also QR', $Q'R$.

54. Conversely, if two chords QQ', RR', which are not parallel, make equal angles with the axis, a circle can be drawn through Q, Q', R', R.

For, if the chords intersect in O, and OE be drawn parallel to the axis and meeting the curve in E, it may be shewn as above that

$$OQ \,.\, OQ' = 4SP \,.\, OE, \text{ and } OR \,.\, OR' = 4SP' \,.\, OE,$$

P and P' being the vertices of the diameters bisecting the chords.

But the tangents at P and P', which are parallel to the chords, are equally inclined to the axis, and therefore SP is equal to SP'.

Hence
$$OQ \,.\, OQ' = OR \,.\, OR',$$

and therefore a circle can be drawn through the points Q, Q', R, R'.

If the two chords are both perpendicular to the axis, it is obvious that a circle can be drawn through their extremities, and this is the only case in which a circle can be drawn through the extremities of parallel chords.

EXAMPLES.

1. Find the locus of the centre of a circle which passes through a given point and touches a given straight line.

2. Draw a tangent to a parabola, making a given angle with the axis.

3. If the tangent at P meet the tangent at the vertex in Y,

$$AY^2 = AS \,.\, AN.$$

4. If the normal at P meet the axis in G, the focus is equidistant from the tangent at P and the straight line through G parallel to the tangent.

5. Given the focus, the position of the axis, and a tangent, construct the parabola.

6. Find the locus of the centre of a circle which touches a given straight line and a given circle.

7. Construct a parabola which has a given focus, and two given tangents.

8. The distance of any point on a parabola from the focus is equal to the length of the ordinate at that point produced to meet the tangent at the end of the latus rectum.

9. PT being the tangent at P, meeting the axis in T, and PN the ordinate, prove that $TY \,.\, TP = TS \,.\, TN$.

10. If SE be the perpendicular from the focus on the normal at P, shew that

$$SE^2 = AN \,.\, SP.$$

11. The locus of the vertices of all parabolas, which have a common focus and a common tangent, is a circle.

12. Having given the focus, the length of the latus rectum, and a tangent, construct the parabola.

13. If PSP' be a focal chord, and PN, $P'N'$ the ordinates, shew that

$$AN \,.\, AN' = AS^2.$$

Shew also that the latus rectum is a mean proportional between the double ordinates.

14. The locus of the middle points of the focal chords of a parabola is another parabola.

15. Shew that in general two parabolas can be drawn having a given straight line for directrix, and passing through two given points on the same side of the line.

16. Pp is a chord perpendicular to the axis, and the perpendicular from p on the tangent at P meets the diameter through P in R; prove that RP is equal to the latus rectum, and find the locus of R.

17. Having given the focus, describe a parabola passing through two given points.

18. The circle on any focal distance as diameter touches the tangent at the vertex.

19. The circle on any focal chord as diameter touches the directrix.

20. A point moves so that its shortest distance from a given circle is equal to its distance from a given diameter of the circle; prove that the locus is a parabola, the focus of which coincides with the centre of the circle.

21. Find the locus of a point which moves so that its shortest distance from a given circle is equal to its distance from a given straight line.

22. The vertex of an isosceles triangle is fixed. The extremities of its base lie on two fixed parallel straight lines. Prove that the base is a tangent to a parabola.

23. Shew that the normal at any point of a parabola is equal to the ordinate through the middle point of the subnormal.

24. If perpendiculars are drawn to the tangents to a parabola where they meet the axis they will be normals to two equal parabolas.

25. PSP' is a focal chord of a parabola. The diameters through P, P' meet the normals at P', P in V, V' respectively. Prove that $PVV'P'$ is a parallelogram.

26. If APC be a sector of a circle, of which the radius CA is fixed, and a circle be described, touching the radii CA, CP, and the arc AP, the locus of the centre of this circle is a parabola.

27. If from the focus S of a parabola, SY, SZ be perpendiculars drawn to the tangent and normal at any point, YZ is parallel to the diameter.

28. Prove that the locus of the foot of the perpendicular from the focus on the normal is a parabola.

29. If PG be the normal, and GL the perpendicular from G upon SP, prove that GL is equal to the ordinate PN.

30. Given the focus, a point P on the curve, and the length of the perpendicular from the focus on the tangent at P, find the vertex.

31. A circle is described on the latus rectum as diameter, and a common tangent QP is drawn to it and the parabola: shew that SP, SQ make equal angles with the latus rectum.

32. G is the foot of the normal at a point P of the parabola, Q is the middle point of SG, and X is the foot of the directrix: prove that

$$QX^2 - QP^2 = 4AS^2.$$

33. If PG the normal at P meet the axis in G, and if PF, PH, lines equally inclined to PG, meet the axis in F and H, the length SG is a mean proportional between SF and SH.

34. A triangle ABC circumscribes a parabola whose focus is S, and through A, B, C, lines are drawn respectively perpendicular to SA, SB, SC; shew that these pass through one point.

35. If PQ be the normal at P meeting the curve in Q, and if the chord PR be drawn so that PR, PQ are equally inclined to the axis, PRQ is a right angle.

36. PN is a semi-ordinate of a parabola, and AM is taken on the other side of the vertex along the axis equal to AN; from any point Q in PN, QR is drawn parallel to the axis meeting the curve in R; prove that the lines MR, AQ will intersect in the parabola.

37. Having given two points of a parabola, the direction of the axis, and the tangent at one of the points, construct the parabola.

38. Having given the vertex of a diameter, and a corresponding double ordinate, construct the parabola.

39. PM is an ordinate of a point P; a straight line parallel to the axis bisects PM, and meets the curve in Q; MQ meets the tangent at the vertex in T; prove that $3AT = 2PM$.

40. AB, CD are two parallel straight lines given in position, and AC is perpendicular to both, A and C being given points; in CD any point Q is taken, and in AQ, produced if necessary, a point P is taken, such that the distance of P from AB is equal to CQ; prove that the locus of P is a parabola.

41. If the tangent and normal at a point P of a parabola meet the tangent at the vertex in K and L respectively, prove that

$$KL^2 : SP^2 :: SP - AS : AS.$$

42. Having given the length of a focal chord, find its position.

43. If the ordinate of a point P bisects the subnormal of a point P', prove that the ordinate of P is equal to the normal of P'.

44. A parabola being traced on a plane, find its axis and vertex.

45. If PV, $P'V'$ be two diameters, and PV', $P'V$ ordinates to these diameters,

$$PV = P'V'.$$

46. If one side of a triangle be parallel to the axis of a parabola, the other sides will be in the ratio of the tangents parallel to them.

47. QVQ' is an ordinate of a diameter PV, and any chord PR meets QQ' in N, and the diameter through Q in L; prove that

$$PL^2 = PN \,.\, PR.$$

48. Describe a parabola passing through three given points, and having its axis parallel to a given line.

49. If AP, AQ be two chords drawn from the vertex at right angles to each other, and PN, QM be ordinates, the latus rectum is a mean proportional between AN and AM.

50. PSp is a focal chord of a parabola; prove that AP, Ap meet the latus rectum in two points whose distances from the focus are equal to the ordinates of p and P respectively.

51. If the straight line AP and the diameter through P meet the double ordinate QMQ' in R and R', prove that

$$RM \,.\, R'M = QM^2.$$

52. A and P are two fixed points. Parabolas are drawn all having their vertices at A, and all passing through P. Prove that the points of intersection of the tangents at P with the tangent and normal at A lie on two fixed circles, one of which is double the size of the other.

53. A variable tangent to a parabola intersects two fixed tangents in the points T and T': shew that the ratio $ST : ST'$ is constant.

54. Through a fixed point on the axis of a parabola a chord PQ is drawn, and a circle of given radius is described through the feet of the ordinates of P and Q. Shew that the locus of its centre is a circle.

55. If SY be the perpendicular on the tangent at P, and if YS be produced to R so that $SR = SY$, shew that PAR is a right angle.

56. If two circles be drawn touching a parabola at the ends of a focal chord, and passing through the focus, shew that they intersect each other orthogonally.

57. PSQ is a focal chord of a parabola, whose vertex is A and focus S, V being the middle point of the chord, shew that

$$PV^2 = AV^2 + 3AS^2.$$

58. QQ' is a focal chord of a parabola. Describe a circle which shall pass through Q, Q' and touch the parabola.

If P be the point of contact and the angle QPQ' a right angle, find the inclination of QP to the axis.

59. Through two fixed points E, F, on the axis of a parabola are drawn two chords PQ, PR meeting the curve in P, Q, R. If QR meet the axis in T, shew that the ratio $TR : TQ$ is constant.

60. A chord PQ is normal to the parabola at P, and the angle PSQ is a right angle. Prove that $SQ = 2SP$, and that the ordinate of P is equal to the latus rectum. Also, if T is the point of intersection of the tangents at P and Q, and if R is the middle point of TQ, prove that the angle TSR is a right angle, and that $ST = 2SR$.

61. A straight line intersects a circle; prove that all the chords of the circle which are bisected by the straight line are tangents to a parabola.

62. If two tangents TP, TQ be drawn to a parabola, the perpendicular SE from the focus on their chord of contact passes through the middle point of their intercept on the tangent at the vertex.

63. From the vertex of a parabola a perpendicular is drawn on the tangent at any point; prove that the locus of its intersection with the diameter through the point is a straight line.

64. If two tangents to a parabola be drawn from any point in its axis, and if any other tangent intersect these two in P and Q, prove that $SP = SQ$.

65. T is a point on the tangent at P, such that the perpendicular from T on SP is of constant length; prove that the locus of T is a parabola.

If the constant length be $2AS$, prove that the vertex of the locus is on the directrix.

66. Given a chord of a parabola in magnitude and position, and the point in which the axis cuts the chord, the locus of the vertex is a circle.

67. If the normal at a point P of a parabola meet the curve in Q, and the tangents at P and Q intersect in T, prove that T and P are equidistant from the directrix.

68. If TP, TQ be tangents to a parabola, such that the chord PQ is normal at P,

$$PQ : PT :: PN : AN,$$

PN and AN being the ordinate and abscissa.

69. If two equal tangents to a parabola be cut by a third tangent, the alternate segments of the two tangents will be equal.

70. If AP be a chord through the vertex, and if PL, perpendicular to AP, and PG, the normal at P, meet the axis in L, G respectively, GL = half the latus rectum.

71. If PSQ be a focal chord, A the vertex, and PA, QA be produced to meet the directrix in P', Q' respectively, then $P'SQ'$ will be a right angle.

72. The tangents at P and Q intersect in T, and the tangent at R intersects TP and TQ in C and D; prove that

$$PC : CT :: CR : RD :: TD : DQ.$$

73. From any point D in the latus rectum of a parabola, a straight line DP is drawn, parallel to the axis, to meet the curve in P; if X be the foot of the directrix, and A the vertex, prove that AD, XP intersect in the parabola.

74. PSp is a focal chord, and upon PS and pS as diameters circles are described; prove that the length of either of their common tangents is a mean proportional between AS and Pp.

75. If AQ be a chord of a parabola through the vertex A, and QR be drawn perpendicular to AQ to meet the axis in R; prove that AR will be equal to the chord through the focus parallel to AQ.

76. If from any point P of a circle, PC be drawn to the centre C, and a chord PQ be drawn parallel to the diameter AB, and bisected in R; shew that the locus of the intersection of CP and AR is a parabola.

77. A circle, the diameter of which is three-fourths of the latus rectum, is described about the vertex A of a parabola as centre; prove that the common chord bisects AS.

78. Shew that straight lines drawn perpendicular to the tangents of a parabola through the points where they meet a given fixed line perpendicular to the axis are in general tangents to a confocal parabola.

79. If QR be a double ordinate, and PD a straight line drawn parallel to the axis from any point P of the curve, and meeting QR in D, prove, from Art. 27, that

$$QD \, . \, RD = 4AS \, . \, PD.$$

80. Prove, by help of the preceding theorem, that, if QQ' be a chord parallel to the tangent at P, QQ' is bisected by PD, and hence determine the locus of the middle point of a series of parallel chords.

81. If a parabola touch the sides of an equilateral triangle, the focal distance of any vertex of the triangle passes through the point of contact of the opposite side.

82. Find the locus of the foci of the parabolas which have a common vertex and a common tangent.

83. From the points where the normals to a parabola meet the axis, lines are drawn perpendicular to the normals: shew that these lines will be tangents to an equal parabola.

84. Inscribe in a given parabola a triangle having its sides parallel to three given straight lines.

85. PNP' is a double ordinate, and through a point of the parabola RQL is drawn perpendicular to PP' and meeting PA, or PA produced in R; prove that

$$PN : NL :: LR : RQ.$$

86. PNP' is a double ordinate, and through R, a point in the tangent at P, RQM is drawn perpendicular to PP' and meeting the curve in Q; prove that

$$QM : QR :: P'M : PM.$$

87. If from the point of contact of a tangent to a parabola, a chord be drawn, and a line parallel to the axis meeting the chord, the tangent, and the curve, shew that this line will be divided by them in the same ratio as it divides the chord.

88. PSp is a focal chord of a parabola, RD is the directrix meeting the axis in D, Q is any point in the curve; prove that if QP, Qp produced meet the directrix in R, r, half the latus rectum will be a mean proportional between DR and Dr.

89. A chord of a parabola is drawn parallel to a given straight line, and on this chord as diameter a circle is described; prove that the distance between the middle points of this chord, and of the chord joining the other two points of intersection of the circle and parabola, will be of constant length.

90. If a circle and a parabola have a common tangent at P, and intersect in Q and R; and if QV, UR be drawn parallel to the axis of the parabola meeting the circle in V and U respectively, then will VU be parallel to the tangent at P.

91. If PV be the diameter through any point P, QV a semi-ordinate, Q' another point in the curve, and $Q'P$ cut QV in R, and $Q'R'$, the diameter through Q', meet QV in R', then

$$VR \,.\, VR' = QV^2.$$

92. PQ, PR are any two chords; PQ meets the diameter through R in the point F, and PR meets the diameter through Q in E; prove that EF is parallel to the tangent at P.

93. If parallel chords be intersected by a diameter, the distances of the points of intersection from the vertex of the diameter are in the ratio of the rectangles contained by the segments of the chords.

94. If tangents be drawn to a parabola from any point P in the latus rectum, and if Q, Q' be the points of contact, the semi-latus rectum is a geometric mean between the ordinates of Q and Q', and the distance of P from the axis is an arithmetic mean between the same ordinates.

95. If A', B', C' be the middle points of the sides of a triangle ABC, and a parabola drawn through A', B', C' meet the sides again in A'', B'', C'', then will the lines AA'', BB'', CC'' be parallel to each other.

96. A circle passing through the focus cuts the parabola in two points. Prove that the angle between the tangents to the circle at those points is four times the angle between the tangents to the parabola at the same points.

97. The locus of the points of intersection of normals at the extremities of focal chords of a parabola is another parabola.

98. Having given the vertex, a tangent, and its point of contact, construct the parabola.

99. PSp is a focal chord of a parabola; shew that the distance of the point of intersection of the normals at P and p from the directrix varies as the rectangle contained by PS, pS.

100. TP, TQ are tangents to a parabola at P and Q, and O is the centre of the circle circumscribing PTQ; prove that TSO is a right angle.

101. P is any point of a parabola whose vertex is A, and through the focus S the chord QSQ' is drawn parallel to AP; PN, QM, $Q'M'$, being perpendicular to the axis, shew that SM is a mean proportional between AM, AN, and that

$$MM' = AP.$$

102. If a circle cut a parabola in four points, two on one side of the axis, and two on the other, the sum of the ordinates of the first two is equal to the sum of the ordinates of the other two points.

Extend this theorem to the case in which three of the points are on one side of the axis and one on the other.

103. The tangents at P and Q meet in T, and TL is the perpendicular from T on the axis; prove that if PN, QM be the ordinates of P and Q,

$$PN \,.\, QM = 4AS \,.\, AL.$$

104. The tangents at P and Q meet in T, and the lines TA, PA, QA, meet the directrix in t, p, and q: prove that

$$tp = tq.$$

105. From a point T tangents TP, TQ are drawn to a parabola, and through T straight lines are drawn parallel to the normals at P and Q; prove that one diagonal of the parallelogram so formed passes through the focus.

106. Through a given point within a parabola draw a chord which shall be divided in a given ratio at that point.

107. ABC is a portion of a parabola bounded by the axis AB and the semi-ordinate BC; find the point P in the semi-ordinate such that if PQ be drawn parallel to the axis to meet the parabola in Q, the sum of BP and PQ shall be the greatest possible.

108. The diameter through a point P of a parabola meets the tangent at the vertex in Z; the normal at P and the focal distance of Z will intersect in a point at the same distance from the tangent at the vertex as P.

109. Given a tangent to a parabola and a point on the curve, shew that the foot of the ordinate of the point of contact of the tangent drawn to the diameter through the given point lies on a fixed straight line.

110. Find a point such that the tangents from it to a parabola and the lines from the focus to the points of contact may form a parallelogram.

111. Two equal parabolas have a common focus; and, from any point in the common tangent, another tangent is drawn to each; prove that these tangents are equidistant from the common focus.

112. Two parabolas have a common axis and vertex, and their concavities turned in opposite directions; the latus rectum of one is eight times that of the other; prove that the portion of a tangent to the former, intercepted between the common tangent and axis, is bisected by the latter.

CHAPTER III.

THE ELLIPSE.

DEF. *An ellipse is the curve traced out by a point which moves in such a manner that its distance from a given point is in a constant ratio of less inequality to its distance from a given straight line.*

Tracing the Curve.

55. Let S be the focus, EX the directrix, and SX the perpendicular on EX from S.

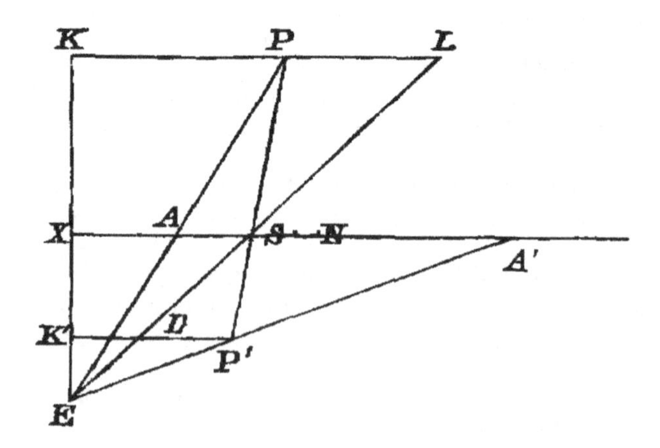

Divide SX at the point A in the given ratio; the point A is the vertex.

From any point E in EX, draw EAP, ESL, and through S draw SP making the angle PSL equal to LSN, and meeting EAP in P.

Through P draw LPK perpendicular to the directrix and meeting ESL in L.

Then the angle $$PSL = LSN = SLP.$$
$$\therefore SP = PL.$$

Also $$PL : PK :: SA : AX.$$

Hence $$SP : PK :: SA : AX,$$

and P is therefore a point in the curve.

Again, in the axis XAN find a point A' such that
$$SA' : A'X :: SA : AX;$$
this point is evidently on the same side of the directrix as the point A, and is another vertex of the curve.

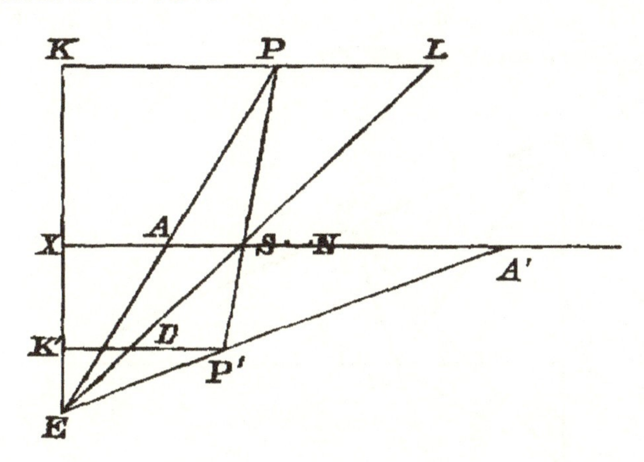

Join EA' meeting PS produced in P', and draw $P'L'K'$ perpendicular to the directrix and meeting ES in L'.

Then $$P'L' : P'K' :: SA' : A'X$$
$$:: SA : AX,$$

and the angle $$SL'P' = L'SA = L'SP;$$
$$\therefore P'L' = SP'.$$

Hence P' is also a point in the curve, and PSP' is a focal chord.

By giving E a series of positions on the directrix we shall obtain a series of focal chords, and we can also, as in Art. (1), find other points of the curve lying in the lines KP, $K'P'$, or in these lines produced.

We can thus find any number of points in the curve.

56. DEF. *The distance AA' is the major axis.*

The middle point C of AA' is called the centre of the ellipse.

If through C the double ordinate BCB' be drawn, BB' is called the minor axis.

Any straight line drawn through the centre, and terminated by the curve, is called a diameter.

The lines ACA', BCB' are called the principal diameters, or, briefly, the axes of the curve.

The line ACA' is also sometimes called the transverse axis, and BCB' the conjugate axis.

57. PROP. I. *If P be any point of an ellipse, and AA' the axis major, and if PA, $A'P$, when produced, meet the directrix in E and F, the distance EF subtends a right angle at the focus.*

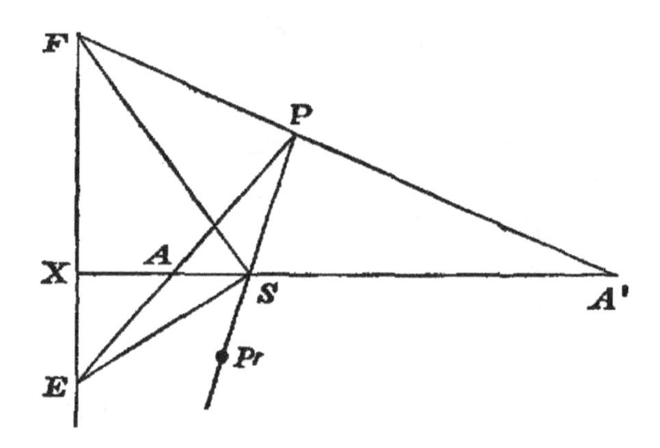

By the theorem of Art. 4, ES bisects the angle ASP', and FS bisects the angle ASP;

$$\therefore ESF \text{ is a right angle.}$$

It will be seen that, since ASA' is a focal chord, this is a particular case of the theorem of Art. 6.

58. PROP. II. *If PN be the ordinate of any point P of an ellipse, ACA' the axis major, and BCB' the axis minor,*

$$PN^2 : AN \, . \, NA' :: BC^2 : AC^2.$$

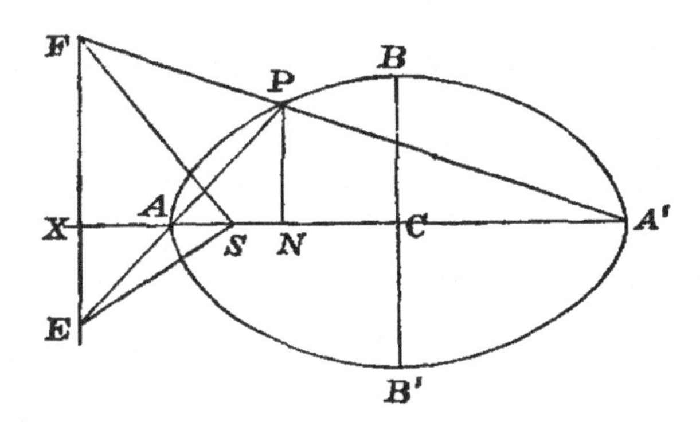

Join PA, $A'P$, and let these lines produced meet the directrix in E and F.

Then $$PN : AN :: EX : AX,$$

and $$PN : A'N :: FX : A'X;$$

$$\therefore PN^2 : AN . NA' :: EX . FX : AX . A'X$$
$$:: SX^2 : AX . A'X,$$

since ESF is a right angle (Prop. 1.); that is, PN^2 is to $AN . NA'$ in a constant ratio.

Hence, taking PN coincident with BC, in which case

$$AN = NA' = AC,$$
$$BC^2 : AC^2 :: SX^2 : AX . A'X,$$

and $$\therefore PN^2 : AN . NA' :: BC^2 : AC^2.$$

This may be also written

$$PN^2 : AC^2 - CN^2 :: BC^2 : AC^2.$$

Cor. If PM be the perpendicular from P on the axis minor,

$$CM = PN, PM = CN,$$

and $$CM^2 : AC^2 - PM^2 :: BC^2 : AC^2.$$

Hence $$AC^2 : AC^2 - PM^2 :: BC^2 : CM^2,$$

and $$\therefore AC^2 : PM^2 :: BC^2 : BC^2 - CM^2,$$

or $$PM^2 : BM . MB' :: AC^2 : BC^2.$$

59. If a point N' be taken on the axis major, between C and A', such that $CN' = CN$, the corresponding ordinate $P'N' = PN$, and therefore it follows that the curve is symmetrical with regard to BCB', and that there is another focus, and another directrix, corresponding to the vertex A'.

60. By help of the theorem of Art. 57, we can give an independent proof of the existence of the other focus and directrix, corresponding to the vertex A'.

In AA' produced take a point X' such that $A'X' = AX$, and in AA' take a point S' such that $A'S' = AS$.

Through X' draw a straight line $eX'f$ perpendicular to the axis, and let EP, FP produced meet this line in e and f. Join eS', and fS'.

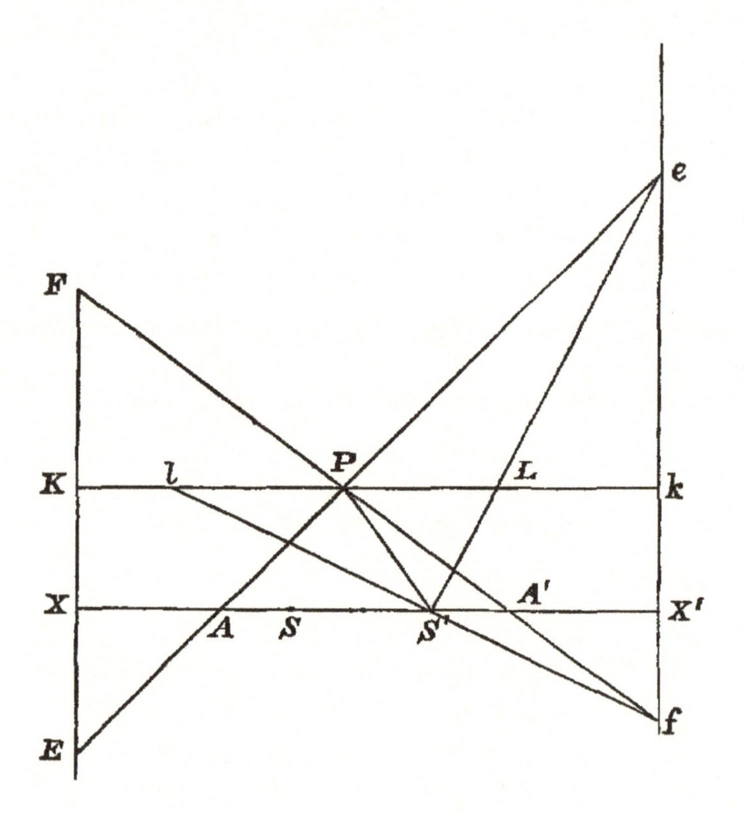

Then
$$eX' : EX :: AX' : AX$$
$$:: A'X : A'X'$$
$$:: FX : fX';$$
$$\therefore eX' \,.\, fX' = EX \,.\, FX = SX^2 = S'X'^2.$$

Hence $eS'f$ is a right angle.

Through P draw KPk parallel to the axis, meeting eS' and fS' produced in L and l.

Then $$PL : Pk :: S'A : AX' :: SA' : A'X,$$

and $$Pl : Pk :: S'A' : A'X' :: SA : AX,$$

$$\therefore PL = Pl.$$

Moreover, $LS'l$ being a right angle,

$$S'P = Pl,$$

$$\therefore S'P : Pk :: S'A' : A'X',$$

and the curve can be described by means of the focus S' and the directrix eX'.

If SA be equal to AX, the point A', and therefore the points S' and X', will be at an infinite distance from S and A.

Hence a parabola is the limiting form of an ellipse, the axis major of which is indefinitely increased in magnitude, while the distance SA remains finite.

61. PROP. III. *If ACA' be the axis major, C the centre, S one of the foci, and X the foot of the directrix,*

$$CS : CA :: CA : CX :: SA : AX,$$

and $$CS : CX :: CS^2 : CA^2.$$

X A S C S' A' X'

For $$S'A : SA :: AX' : AX$$
$$:: A'X : AX;$$
$$\therefore SS' : SA :: AA' : AX,$$

or $$CS : CA :: SA : AX.$$

Again, $$SA' : SA :: AX' : AX;$$
$$\therefore AA' : SA :: XX' : AX,$$

or $$CA : CX :: SA : AX;$$
$$\therefore CS : CA :: CA : CX,$$

or $$CS \,.\, CX = CA^2.$$

Also $$CS : CX :: CS^2 : CS \,.\, CX$$
$$:: CS^2 : CA^2.$$

62. Prop. IV. *If S be a focus, and B an extremity of the axis minor,* $SB = AC$ and $BC^2 = AS \,.\, SA'$.

For, joining SB in the figure of Art. 58,

$$SB : CX :: SA : AX$$
$$:: CA : CX,$$

by the previous Article,

$$\therefore SB = CA.$$

Also $$BC^2 = SB^2 - SC^2 = AC^2 - SC^2$$
$$= AS \,.\, SA'.$$

63. Prop. V. *The semi-latus rectum SR is a third proportional to AC and BC.*

For, Prop. II.,

$$SR^2 : AS \,.\, SA' :: BC^2 : AC^2;$$
$$\therefore SR^2 : BC^2 :: BC^2 : AC^2,$$

or $$SR : BC :: BC : AC.$$

Cor. Since $$SR : SX :: SA : AX$$
$$:: SC : AC,$$

it follows that $SX \,.\, SC = SR \,.\, AC = BC^2$;
and hence also, since $SC \,.\, CX = AC^2$, that

$$SX : CX :: BC^2 : AC^2.$$

64. Prop. VI. *The sum of the focal distances of any point is equal to the axis major.*

Let PN be the ordinate of a point P (Fig. Art. 60), then

$$S'P : SP :: NX' : NX;$$
$$\therefore S'P + SP : SP :: XX' : NX,$$

or $$S'P + SP : XX' :: SP : NX$$
$$:: SA : AX$$
$$:: AA' : XX';$$
$$\therefore S'P + SP = AA'$$

COR. Since $SP : NX :: SA : AX$

$$:: AC : CX;$$

$$\therefore AC : SP :: CX : NX,$$

$$AC - SP : SP :: CN : NX,$$

and $AC - SP : CN :: SA : AX.$

Also, $AC - SP = S'P - AC;$

$$\therefore S'P - AC : CN :: SA : AX.$$

Hence, $S'P - SP : 2CN :: SA : AX.$

Mechanical Construction of the Ellipse.

65. Fasten the ends of a piece of thread to two pins fixed on a board, and trace a curve on the board with a pencil pressed against the thread so as to keep it stretched; the curve traced out will be an ellipse, having its foci at the points where the pins are fixed, and having its major axis equal to the length of the thread.

66. PROP. VII. *The sum of the distances of a point from the foci of an ellipse is greater or less than the major axis according as the point is outside or inside the ellipse.*

If the point be without the ellipse, join SQ, $S'Q$, and take a point P on the intercepted arc of the curve.

Then P is within the triangle SQS' and therefore, joining SP, $S'P$,

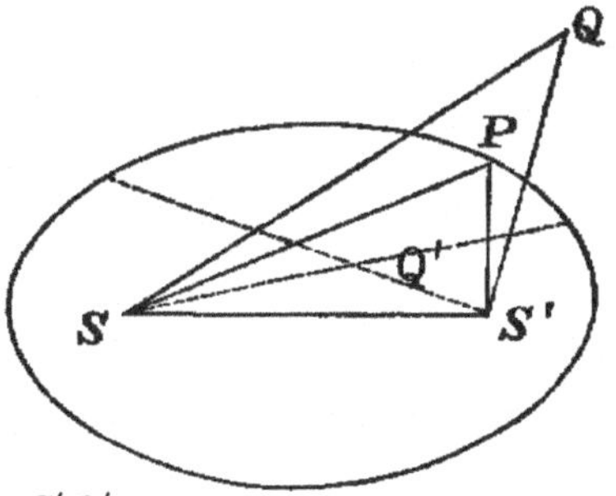

$SQ + S'Q > SP + S'P$, Euclid I. 21,

i.e. $SQ + S'Q > AA'$.

If Q' be within the ellipse, let SQ', $S'Q'$ produced meet the curve and take a point P on the intercepted arc.

Then Q' is within the triangle SPS', and

$$\therefore SP + S'P > SQ' + S'Q',$$

i.e. $SQ' + S'Q' < AA'$.

67. DEF. *The circle described on the axis major as diameter is called the auxiliary circle.*

PROP. VIII. *If the ordinate NP of an ellipse be produced to meet the auxiliary circle in Q,*

$$PN : QN :: BC : AC.$$

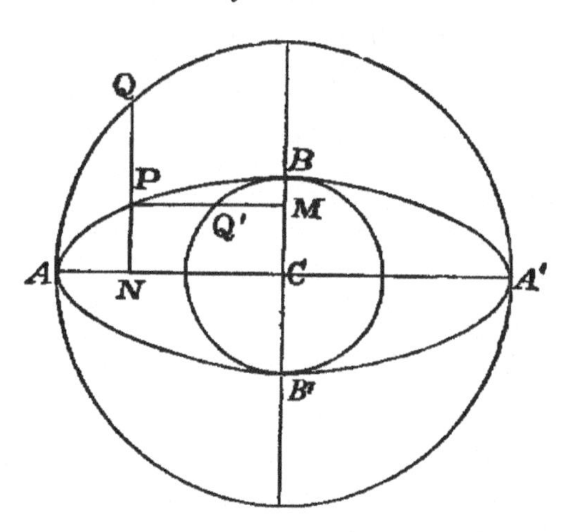

For (Art. 58)

$$PN^2 : AN \,.\, NA' :: BC^2 : AC^2,$$

and, by a property of the circle,

$$QN^2 = AN \,.\, NA'; \therefore PN : QN :: BC : AC.$$

COR. Similarly, if PM, the perpendicular on BB', meet in Q' the circle described on BB' as diameter,

$$PM : Q'M :: AC : BC.$$

For $$PM^2 : BM \,.\, MB' :: AC^2 : BC^2,$$

and $$BM \,.\, MB' = Q'M^2.$$

Properties of the Tangent and Normal.

68. PROP. IX. *The normal at any point bisects the angle between the focal distances of that point, and the tangent is equally inclined to the focal distances.*

Let the normal at P meet the axis in G; then (Art. 18)

$$SG : SP :: SA : AX,$$

and $$S'G : S'P :: SA : AX.$$

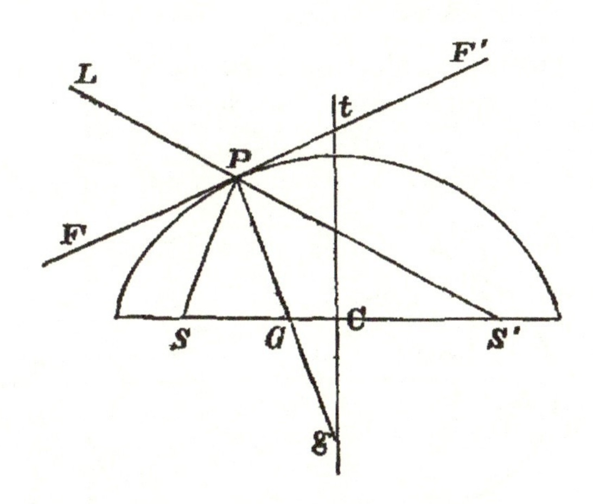

Hence $$SG : S'G :: SP : S'P,$$

and therefore the angle SPS' is bisected by PG.

Also FPF' being the tangent, and GPF, GPF' being right angles, it follows that the angles SPF, $S'PF'$ are equal, or that the tangent is equally inclined to the focal distances.

Hence if $S'P$ be produced to L, the tangent bisects the angle SPL.

COR. If a circle be described about the triangle SPS', its centre will lie in BCB', which bisects SS' at right angles; and since the angles SPG, $S'PG$ are equal, and equal angles stand upon equal arcs, the point g, in which PG produced meets the minor axis, is a point in the circle.

Also, if the tangent meet the minor axis in t, the point t is on the same circle, since gPt is a right angle.

Hence, *Any point P of an ellipse, the two foci, and the points of intersection of the tangent and normal at P with the minor axis are concyclic.*

69. PROP. X. *Every diameter is bisected at the centre, and the tangents at the ends of a diameter are parallel.*

Let PCp be a diameter, PN, pn the ordinates of P and p.

Then $$CN^2 : Cn^2 :: PN^2 : pn^2$$
$$:: AC^2 - CN^2 : AC^2 - Cn^2 \text{ (Art. 58)};$$
$$\therefore CN^2 : AC^2 :: Cn^2 : AC^2.$$

Hence $$CN = Cn \text{ and } \therefore CP = Cp.$$

Draw the focal distances; then, since Pp and SS' bisect each other in C, the figure $SPS'p$ is a parallelogram, and the angle

$$SPS' = SpS'.$$

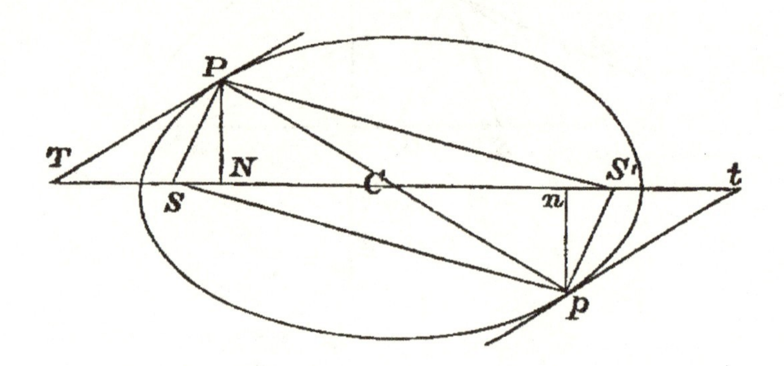

But the tangents PT, pt are equally inclined to the focal distances;

$$\therefore \text{ the angle } SPT = S'pt,$$

and, adding the equal angles CPS, CpS',

$$CPT = Cpt;$$

$$\therefore PT \text{ and } pt \text{ are parallel.}$$

COR. Since Sp and $S'p$ are equally inclined to the tangent at p, it follows that SP and Sp make equal angles with the tangents at P and p.

70. PROP. XI. *The perpendiculars from the foci on any tangent meet the tangent on the auxiliary circle, and the semi-minor axis is a mean proportional between their lengths.*

Let SY, $S'Y'$ be the perpendiculars; join $S'P$, and let SY, $S'P$ produced meet in L.

The angles SPY, YPL being equal, and PY being common, the triangles SPY, YPL are equal in all respects;

$$\therefore PL = SP,\ SY = YL,$$

and $$S'L = S'P + PL = S'P + SP = AA'.$$

Join CY, then C being the middle point of SS', and Y of SL, CY is parallel to $S'L$,

and $$\therefore S'L = 2CY.$$

Hence $CY = AC$, and Y is a point on the auxiliary circle.

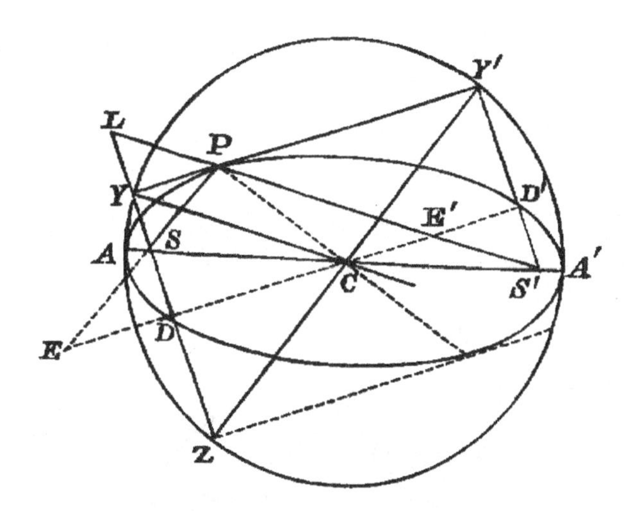

Similarly by producing SP, $S'Y'$ it may be shewn that Y' is also on the auxiliary circle.

Let YS produced meet the circle in Z, and join $Y'Z$; then $Y'YZ$ being a right angle, $Y'Z$ is a diameter and passes through C.

Hence the triangles SCZ, $S'CY'$ are equal, and

$$SY\,.\,S'Y' = SY\,.\,SZ = AS\,.\,SA' = BC^2.$$

Cor. (1). If P' be the other extremity of the diameter through P, the tangent at P' is parallel to PY, and therefore Z is the foot of the perpendicular from S on the tangent at P'.

Cor. (2). If the diameter DCD', drawn parallel to the tangent at P, meet SP, $S'P$ in E and E', $PECY'$ is a parallelogram, for CY' is parallel to SP, and CE to PY';

$$\therefore PE = CY' = AC; \text{ and similarly } PE' = CY = AC.$$

Cor. (3). Any diameter parallel to the focal distance of a point meets the tangent at the point on the auxiliary circle.

71. Prop. XII. *To draw tangents from a given point to an ellipse.*

For this purpose we may employ the general construction of Art. (17), or the following.

Let Q be the given point; upon SQ as diameter describe a circle cutting the auxiliary circle in Y and Y'; YQ and $Y'Q$ will be the required tangents.

Producing SY to L so that $YL = SY$, join $S'L$ cutting the line YQ in P.

The triangles SPY, LPY are equal in all respects, since $SY = YL$ and PY is common and perpendicular to SL;

$$\therefore SP = PL, \text{ and } S'L = S'P + PL = S'P + SP;$$

but, joining CY, $S'L = 2CY = 2AC$;

$$\therefore SP + S'P = 2AC,$$

and P is therefore a point on the ellipse.

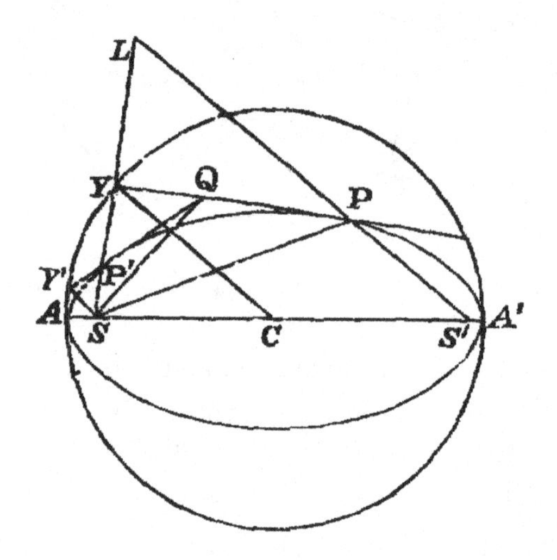

Also the angle $SPY = YPL$,

and $\therefore QP$ is the tangent at P.

A similar construction will give the point of contact of the other tangent QP'.

Referring to Art. 35 it will be seen that the construction is the same as that given for the parabola, the ultimate form of the circle being, for the parabola, the tangent at the vertex.

72. PROP. XIII. *If two tangents be drawn to an ellipse from an external point, they are equally inclined to the focal distances of that point.*

Let QP, QP' be the tangents, SY, $S'Y'$, SZ, $S'Z'$ the perpendiculars from the foci on the tangents; join YZ, $Y'Z'$.

Then (Art. 70)

$$SY \,.\, S'Y' = SZ \,.\, S'Z';$$

$$\therefore SY : SZ :: S'Z' : S'Y'.$$

The points S, Y, Q, Z being concyclic, the angles YSZ, YQZ are supplementary; and similarly, $Z'S'Y'$, $Z'QY'$ are supplementary.

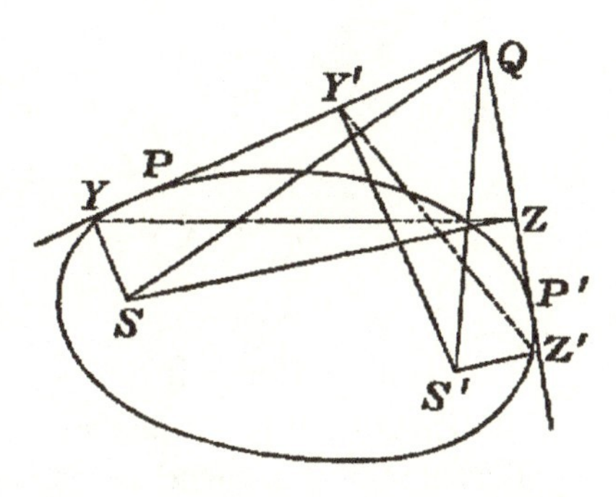

Therefore the angle $YSZ = Z'S'Y'$ and the triangles YSZ, $Z'S'Y'$ are similar.

Therefore the angle $SQP = SZY = S'Y'Z' = S'QP'$.

73. DEF. *Ellipses which have the same foci are called confocal ellipses.*

If Q be a point in a confocal ellipse the normal at Q bisects the angle SQS' and therefore bisects the angle PQP'.

Hence, *If from any point of an ellipse tangents are drawn to a confocal ellipse, these tangents are equally inclined to the normal at the point.*

By reference to the remark of Art. 41, it will be seen that this theorem includes that of Art. 41 as a particular case.

74. PROP. XIV. If PT the tangent at P meet the axis major in T, and PN be the ordinate,

$$CN \,.\, CT = AC^2.$$

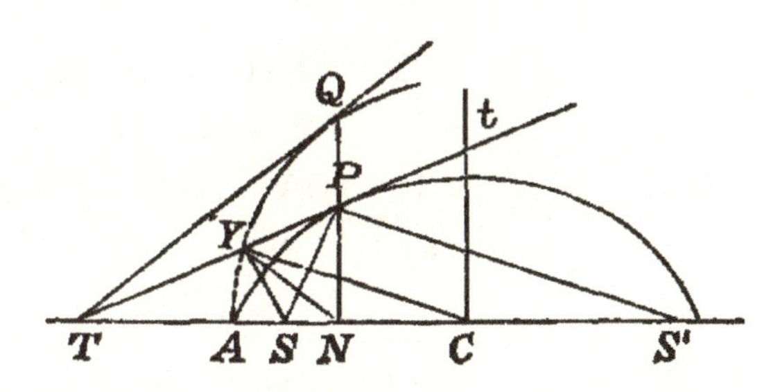

Draw the focal distances SP, $S'P$, and the perpendicular SY on the tangent, and join NY, CY.

Then, as in Art. 70, CY is parallel to $S'P$; therefore the angle

$$CYP = S'Pt = SPY$$
$$= SNY,$$

since S, Y, P, N are concyclic.

Hence $$CYT = CNY,$$

and the triangles CYT, CNY are equiangular.

Therefore $$CN : CY :: CY : CT$$

or $$CN . CT = CY^2 = AC^2.$$

COR. (1). $CN . NT = CN . CT - CN^2 = AC^2 - CN^2$

$$= AN . NA'.$$

COR. (2). Hence it follows that *tangents at the extremities of a common ordinate of an ellipse and its auxiliary circle meet the axis in the same point.*

For, if NP produced meet the auxiliary circle in Q, and the tangent at Q meet the axis in T',

$$CN . NT' = CQ^2 = AC^2,$$

therefore T' coincides with T.

And more generally it is evident that, *If any number of ellipses be described having the same major axis, and an ordinate be drawn cutting the ellipses, the tangents at the points of section will all meet the common axis in the same point.*

75. PROP. XV. *If the tangent at P meet the axis minor in t, and PN be the ordinate,*

$$Ct . PN = BC^2.$$

For, $$Ct : PN :: CT : NT \text{ (Fig. Art. 74)},$$

$$\therefore Ct . PN : PN^2 :: CT . CN : CN . NT$$

$$:: AC^2 : AN . NA' \text{ (Cor. 1, Art. 74)},$$

$$:: BC^2 : PN^2.$$

$$\therefore Ct . PN = BC^2.$$

76. PROP. XVI. *If the tangent and normal at P meet the axis major in T and G,*

$$CG . CT = SC^2.$$

The triangles CGg, CTt, in the figure of the next article, being similar,

$$CG : Cg :: Ct : CT,$$

$$\therefore CG . CT = Cg . Ct.$$

But, since t, S, g, S' are concyclic (Cor. Art. 68),

$$Cg . Ct = SC . CS' = SC^2;$$

$$\therefore CG . CT = SC^2.$$

COR. Since $CN \,.\, CT = AC^2$, and $PN \,.\, Ct = BC^2$,

$$CG : CN :: SC^2 : AC^2$$

and

$$Cg : PN :: SC^2 : BC^2.$$

We hence see that

$$NG : CN :: BC^2 : AC^2.$$

77. PROP. XVII. *If the normal at P meet the axes in G and g, and the diameter parallel to the tangent at P in F,*

$$PF \,.\, PG = BC^2, \text{ and } PF \,.\, Pg = AC^2.$$

Let PN, PM, perpendiculars on the axes, meet the diameter in K and L, and let the tangent at P meet the axes in T and t.

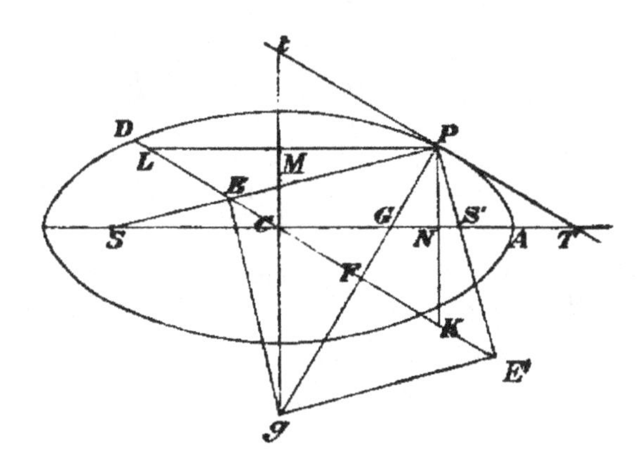

Then, since G, F, K, N are concyclic,

$$PF \,.\, PG = PN \,.\, PK = PN \,.\, Ct = BC^2.$$

Similarly, since L, M, F, g are concyclic,

$$PF \,.\, Pg = PM \,.\, PL = CN \,.\, CT = AC^2.$$

COR. If SP, $S'P$ meet the diameter DCD' parallel to the tangent at P in E and E',

$$PE = AC \text{ (Cor. 2, Art. 70)};$$

$$\therefore PF \,.\, Pg = PE^2 = PE'^2,$$

and hence it follows that the angles PEg, $PE'g$ are right angles.

78. Prop. XVIII. *If PCp be a diameter, QVQ' a chord parallel to the tangent at P and meeting Pp in V, and if the tangent at Q meet pP produced in T,*

$$CV\,.\,CT = CP^2.$$

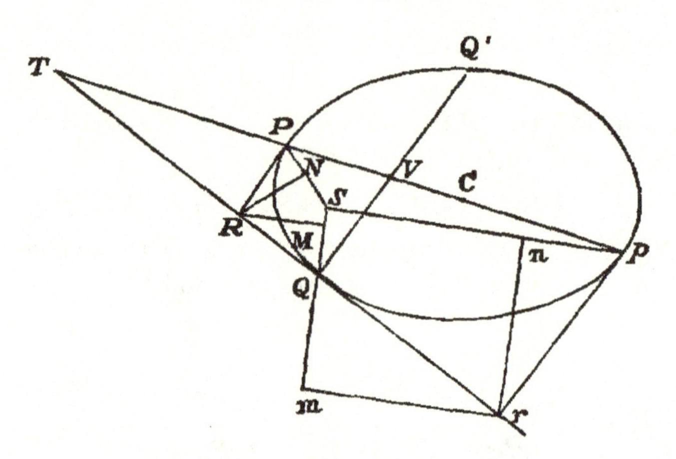

Let TQ meet the tangents at P and p in R and r, and S being a focus, join SP, SQ, Sp.

Let fall perpendiculars RN, RM, rn, rm upon these focal distances; then, since the angle $SPR = Spr$ (Cor. Art. 69),

$$RP : rp :: RN : rn$$
$$:: RM : rm \text{ (Cor. Art. 15)},$$
$$:: RQ : rQ;$$
$$:: PV : Vp.$$

Hence $$TP : Tp :: PV : Vp,$$

or $$CT - CP : CT + CP :: CP - CV : CP + CV;$$

$$\therefore CT : CP :: CP : CV,$$

or $$CT\,.\,CV = CP^2.$$

Cor. 1. Hence, since CV and CP are the same for the point Q', the tangent at Q' passes through T.

Cor. 2. Since $Tp : TP :: pV : VP$, it follows that $TPVp$ is harmonically divided.

It will be seen in a subsequent chapter that this is a particular case of a general theorem.

Properties of Conjugate Diameters.

79. Prop. XIX. *A diameter bisects all chords parallel to the tangents at its extremities.*

We have shewn in Art. 21, that, if QQ' be a chord of a conic, TQ, TQ' the tangents at Q, Q', and EPE' a tangent parallel to QQ', the length EE' is bisected at P.

Draw the diameter PCp; the tangent epe' at p is parallel to EPE' (Art. 69), and is therefore parallel to QQ'.

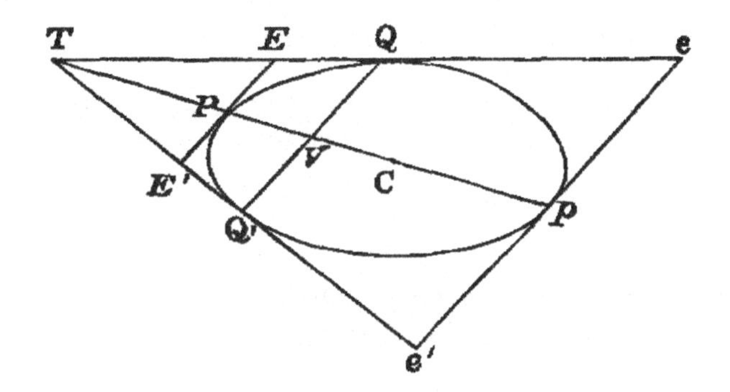

Hence $ep = pe'$, and P, p being the middle points of the parallels ee', EE' the line Pp passes through T, and moreover bisects QQ'.

Similarly, if any other chord qq' be drawn parallel to QQ' the tangents at q and q' will meet in pP produced, and qq' will be bisected by pP.

Cor. Hence, if QQ', qq' be two chords parallel to the tangent at P, the chords Qq, $Q'q'$ will meet in CP or CP produced.

80. Def. *The diameter DCd, drawn parallel to the tangent at P, is said to be conjugate to PCp.*

A diameter therefore bisects all chords parallel to its conjugate.

Prop. XX. *If the diameter DCd be conjugate to PCp, then will PCp be conjugate to DCd.*

Let the chord QVq be parallel to DCd, and therefore bisected by PC, and draw the diameter qCR.

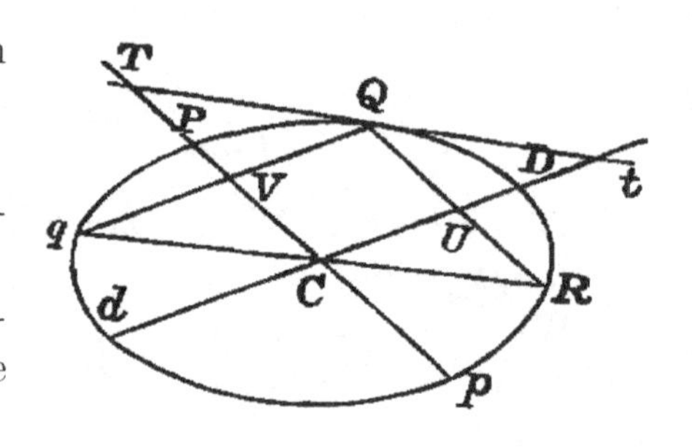

Join QR meeting CD in U; then $RC = Cq$, and $QV = Vq$;

$\therefore QR$ is parallel to CP.

Also $QU : UR :: qC : CR$, and therefore $QU = UR$.

That is, CD bisects the chords parallel to PCp; therefore PCp is conjugate to DCd.

DEF. *Chords drawn from the extremities of a diameter to any point of the ellipse are called supplemental chords.*

Thus qQ, RQ are supplemental chords, and hence it appears that supplemental chords are parallel to conjugate diameters.

DEF. *A line QV drawn from a point Q of an ellipse, parallel to the tangent at P and terminated by the diameter PCp, is called an ordinate of that diameter, and QVq is the double ordinate if QV produced meet the curve in q.*

81. *Any diameter is a mean proportional between the transverse axis and the focal chord parallel to the diameter.*

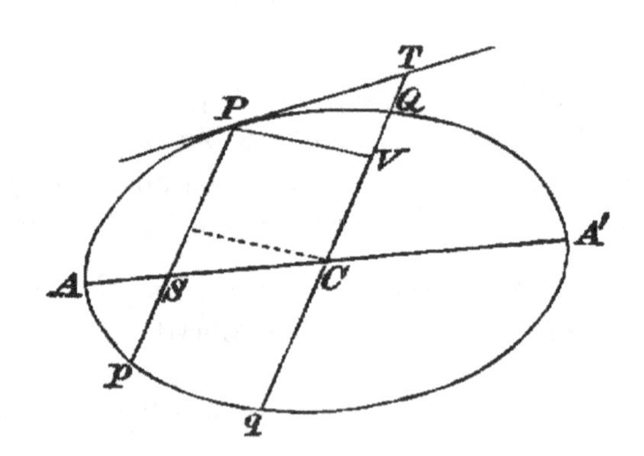

From Art. 70, it appears that if CQT parallel to SP meet in T the tangent at P,

$$CT = AC.$$

Draw PV parallel to the tangent at Q;

$$\text{then } CQ^2 = CV \,.\, CT = CV \,.\, AC;$$

but the diameter through C parallel to the tangent at Q bisects Pp (Art. 80),

$$\text{so that } Pp = 2CV;$$
$$\therefore Qq^2 = Pp \,.\, AA'.$$

82. PROP. XXI. *If PCp, DCd be conjugate diameters, and QV an ordinate of Pp,*

$$QV^2 : PV \,.\, Vp :: CD^2 : CP^2.$$

Let the tangent at Q (Fig. Art. 80) meet CP, CD produced in T and t, and draw QU parallel to CP and meeting CD in U.

Then $$CP^2 = CV \,.\, CT,$$
and $$CD^2 = CU \,.\, Ct = QV \,.\, Ct;$$
$$\therefore CD^2 : CP^2 :: QV \,.\, Ct : CV \,.\, CT$$
$$:: QV^2 : CV \,.\, VT,$$
and $$CV \,.\, VT = CV \,.\, CT - CV^2 = CP^2 - CV^2$$
$$= PV \,.\, Vp,$$
$$\therefore CD^2 : CP^2 :: QV^2 : PV \,.\, Vp.$$

83. PROP. XXII. *If ACA', BCB' be a pair of conjugate diameters, PCP', DCD' another pair, and if PN, DM be ordinates of ACA',*

$$CN^2 = AM \,.\, MA', \quad CM^2 = AN \,.\, NA',$$
$$CM : PN :: AC : BC,$$
and $$DM : CN :: BC : AC.$$

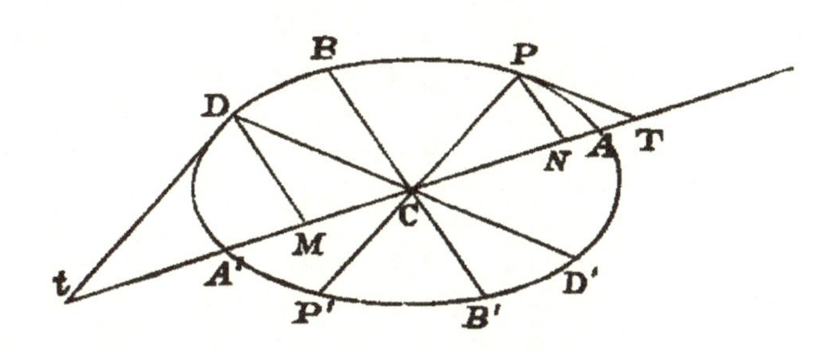

Let the tangents at P and D meet ACA' in T and t.

Then $$CN \,.\, CT = AC^2 = CM \,.\, Ct;$$
hence
$$\begin{aligned} CM : CN &:: CT : Ct \\ &:: PT : CD \\ &:: PN : DM \\ &:: CN : Mt, \end{aligned}$$
$$\therefore CN^2 = CM \,.\, Mt = AC^2 - CM^2 = AM \,.\, MA',$$
and similarly, $$CM^2 = AN \,.\, NA'.$$

Also $$DM^2 : AM \,.\, MA' :: BC^2 : AC^2,$$
$$\therefore DM : CN :: BC : AC,$$
and similarly $$CM : PN :: AC : BC.$$

Cor. We have shewn in the course of the proof that
$$CN^2 + CM^2 = AC^2.$$

By similar reasoning it appears that if Pn, Dm, be ordinates of BCB',
$$Cn^2 + Cm^2 = BC^2;$$
$$\therefore PN^2 + DM^2 = BC^2.$$

It should be noticed that these relations are shewn to be true when ACA', BCB' are any conjugate diameters, including of course the principal axes.

84. Prop. XXIII. If CP, CD be conjugate semi-diameters, and AC, BC the principal semi-diameters,
$$CP^2 + CD^2 = AC^2 + BC^2.$$

From the preceding article,
$$CN^2 + CM^2 = AC^2,$$
and $$PN^2 + DM^2 = BC^2;$$
also ACB being in this case a right angle,
$$PN^2 + CN^2 = CP^2,$$
and $$DM^2 + CM^2 = CD^2,$$
$$\therefore CP^2 + CD^2 = AC^2 + BC^2.$$

85. Def. *If the ordinate NP of a point, when produced, meets the auxiliary circle in Q, the angle ACQ is called the eccentric angle of the point P.*

Prop. XXIV. *If CP, CD be conjugate semi-diameters, the difference between the eccentric angles of P and D is a right angle.*

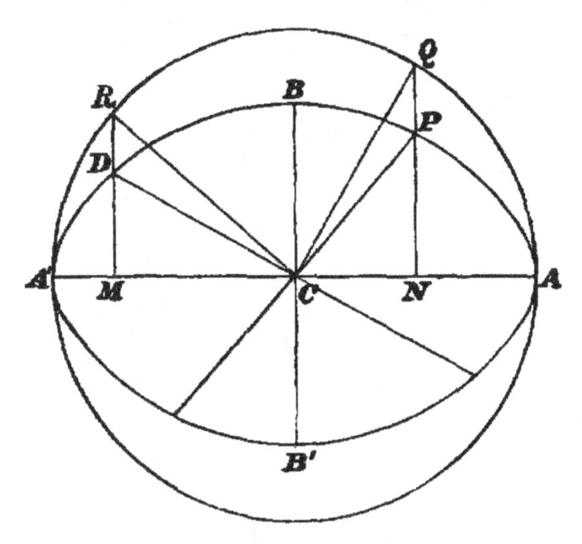

From Art. 67, $RM : DM :: AC : BC$

and, from Art. 83, $CN : DM :: AC : BC$

$$\therefore RM = CN, \text{ and similarly, } QN = CM.$$

$\therefore$ The triangles QCN, CRM are equal, and the angles QCN, RCM are complementary.

$$\therefore QCR \text{ is a right angle.}$$

86. Prop. XXV. *If the normal at P meet the principal axes in G and g,*

$$PG : CD :: BC : AC,$$

and

$$Pg : CD :: AC : BC.$$

For, the triangles DCM, PGN being similar,

$$PG : CD :: PN : CM$$
$$:: BC : AC.$$

So also Pgn and DCM are similar, and

$$Pg : CD :: Pn : DM$$
$$:: AC : BC.$$

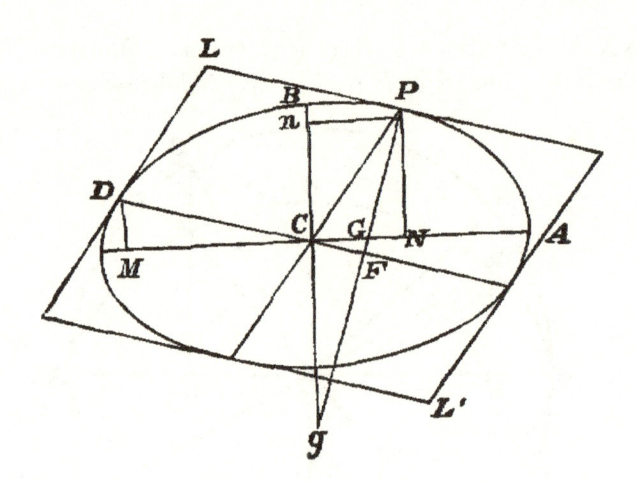

Hence it follows that

$$PG \,.\, Pg = CD^2.$$

87. PROP. XXVI. *The parallelogram formed by the tangents at the ends of conjugate diameters is equal to the rectangle contained by the principal axes.*

For, taking the preceding figure,

$$PG : BC :: CD : AC;$$

but $$PG : BC :: BC : PF \text{ (Art. 77)},$$

$$\therefore CD : AC :: BC : PF,$$

and $$CD \,.\, PF = AC \,.\, BC,$$

whence the theorem stated.

88. PROP. XXVII. *If SP, $S'P$ be the focal distances of P, and CD be conjugate to CP,*

$$SP \,.\, S'P = CD^2,$$

and $$SY : SP :: BC : CD.$$

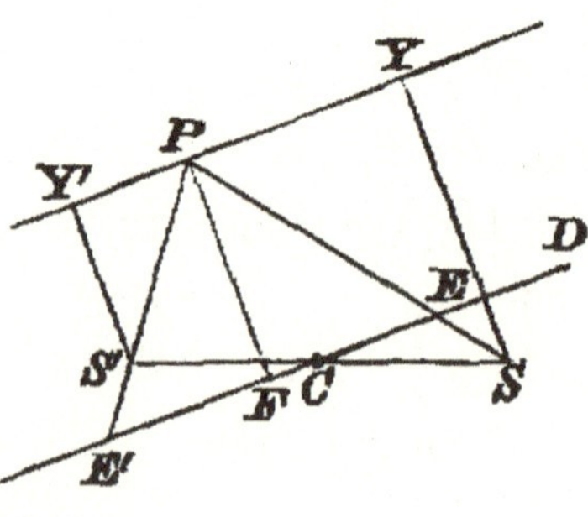

Let CD meet SP, $S'P$ in E and E', and the normal at P in F; then SPY, PEF, and $S'PY'$ are similar triangles;

$$\therefore SP : SY :: PE : PF,$$

and $$S'P : S'Y' :: PE : PF;$$

$$\therefore SP \,.\, S'P : SY \,.\, S'Y' :: PE^2 : PF^2$$
$$:: AC^2 : PF^2$$
$$:: CD^2 : BC^2 \text{ (Art. 87);}$$
$$\therefore SP \,.\, S'P = CD^2.$$

Also $$SY : SP :: PF : PE :: PF : AC,$$
$$\therefore SY : SP :: BC : CD.$$

89. PROP. XXVIII. *If the tangent at P meet a pair of conjugate diameters in T and T', and CD be conjugate to CP,*

$$PT \,.\, PT' = CD^2.$$

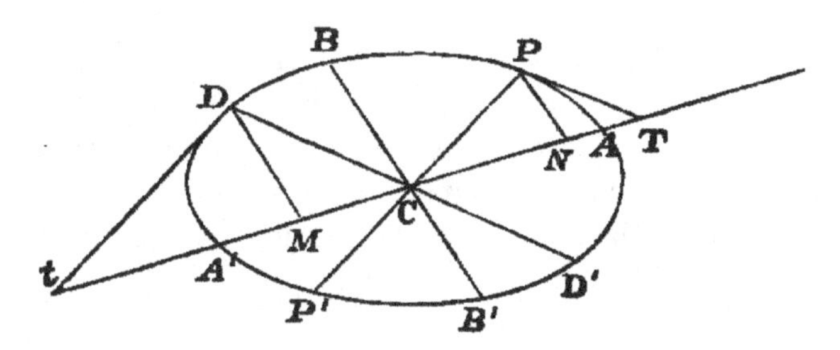

From the figure

$$PT : PN :: CD : DM;$$

and, if TP produced meet CB in T',

$$PT' : CN :: CD : CM;$$
$$\therefore PT \,.\, PT' : PN \,.\, CN :: CD^2 : DM \,.\, CM.$$

But $$PN \,.\, CN = DM \,.\, CM \text{ (Art. 83),}$$
$$\therefore PT \,.\, PT' = CD^2.$$

COR. Let TQU be the tangent at the other end of the chord PNQ, meeting CB' produced in U; and let CE be the semi-diameter parallel to TQ.

Then $$TP : TQ :: PT' : QU,$$
$$\therefore TP^2 : TQ^2 :: PT \,.\, PT' : QT \,.\, QU$$
$$:: CD^2 : CE^2,$$

that is, *the two tangents drawn from any point are in the ratio of the parallel diameters.*

In a similar manner it can be shewn that, if the tangent at P meet the tangents at the ends of a diameter ACA' in T and T',

$$PT \,.\, PT' = CD^2,$$

CD being conjugate to CP,

and $$AT \,.\, A'T' = CB^2,$$

CB being conjugate to ACA'.

90. *Equi-conjugate diameters.*

PROP. XXIX. *The diagonals of the rectangle formed by the principal axes are equal and conjugate diameters.*

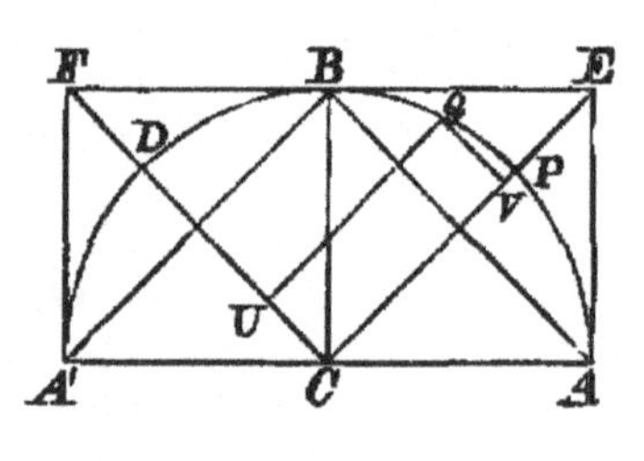

For, joining AB, $A'B$, these lines are parallel to the diagonals CF, CE; and, AB, $A'B$ being supplemental chords, it follows that CD, CP are conjugate to each other. Moreover, they are equally inclined to the axes, and are therefore of equal length.

COR. 1. If QV, QU be drawn parallel to the equi-conjugate diameters, meeting them in V and U,

$$QV^2 : CP^2 - CV^2 :: CD^2 : CP^2;$$

$$\therefore\ QV^2 = CP^2 - CV^2 = PV \,.\, VP',$$

if P' be the other end of the diameter PCP'.

Hence $$QV^2 + QU^2 = CP^2.$$

COR. 2. $$CP^2 + CD^2 = AC^2 + BC^2 \text{ (Art. 84)};$$

$$\therefore\ 2CP^2 = AC^2 + BC^2.$$

91. PROP. XXX. *Pairs of tangents at right angles to each other intersect on a fixed circle.*

The two tangents being TP, TP', let $S'P$ produced meet SY the perpendicular on TP in K.

Then the angle $PTK = STP = S'TP'$;

$$\therefore\ S'TK \text{ is a right angle.}$$

Hence $$4AC^2 = S'K^2 = S'T^2 + TK^2$$
$$= S'T^2 + ST^2$$
$$= 2CT^2 + 2CS^2 \text{ (Euclid, II. 12 and 13)};$$

$$\therefore\ CT^2 = AC^2 + BC^2,$$

and T lies on a fixed circle, of which C is the centre.

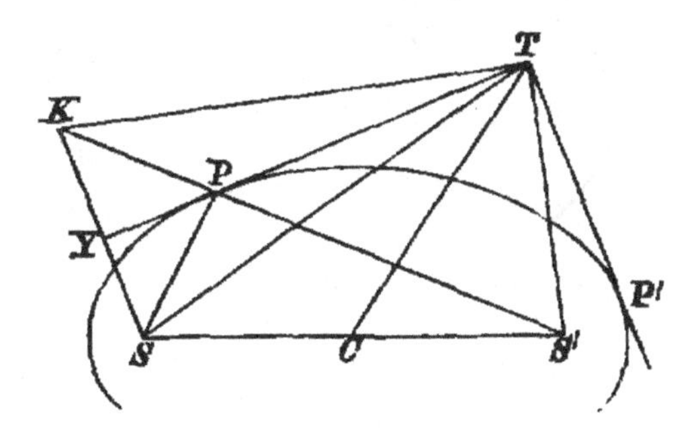

This circle is called the *Director Circle* of the Ellipse, and it will be seen that when the ellipse, by the elongation of SC from S is transformed into a parabola, the director circle merges into the directrix of the parabola.

Cor. If XQ is the tangent to the director circle from the foot of the directrix,

$$XQ^2 = CX^2 - CQ^2 = CX^2 - CA^2 - CB^2$$
$$= CX^2 - SC \,.\, CX - SC \,.\, SX \text{ (Arts. 61 and 63)},$$
$$= CX \,.\, SX - SC \,.\, SX = SX^2.$$
$$\therefore\ XQ = SX,$$

and hence it follows that *the directrix is the radical axis of the director circle and of a point circle at the focus.*

92. Prop. XXXI. *The rectangles contained by the segments of any two chords which intersect each other are in the ratio of the squares of the parallel diameters.*

Through any point O in a chord OQQ' draw the diameter ORR', and let CD be parallel to QQ', and CP conjugate to CD, bisecting QQ' in V.

Draw RU parallel to CD.

Then $$CD^2 - RU^2 : CU^2 :: CD^2 : CP^2 \text{ (Art. 82)},$$
$$:: CD^2 - QV^2 : CV^2.$$

But $$RU^2 : CU^2 :: OV^2 : CV^2;$$
$$\therefore\ CD^2 : CU^2 :: CD^2 + OV^2 - QV^2 : CV^2$$

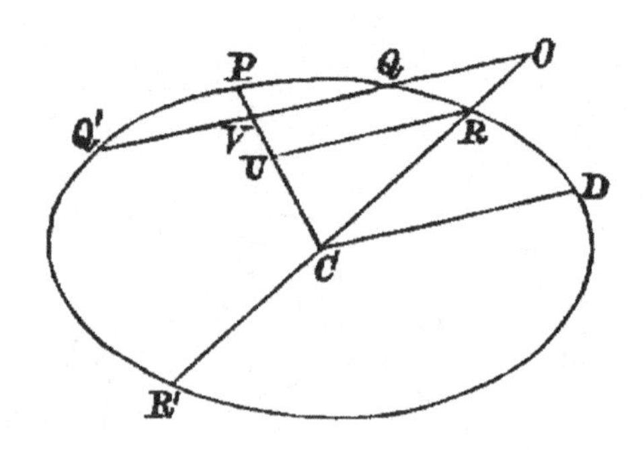

or $$CD^2 : CD^2 + OV^2 - QV^2 :: CU^2 : CV^2$$
$$:: CR^2 : CO^2;$$
$$\therefore\ CD^2 : OV^2 - QV^2 :: CR^2 : CO^2 - CR^2,$$
or $$CD^2 : OQ\,.\,OQ' :: CR^2 : OR\,.\,OR'.$$

Similarly, if Oqq' be any other chord through O, and Cd the parallel semi-diameter,

$$Cd^2 : Oq\,.\,Oq' :: CR^2 : OR\,.\,OR';$$
$$\therefore\ OQ\,.\,OQ' : Oq\,.\,Oq' :: CD^2 : Cd^2.$$

This may otherwise be expressed thus,

The ratio of the rectangles of the segments depends only on the directions in which they are drawn.

The proof is the same if the point O be within the ellipse.

93. PROP. XXXII. *If a circle intersect an ellipse in four points, the several pairs of the chords of intersection are equally inclined to the axes.*

For if QQ', qq' be a pair of the chords of intersection, and if these meet in O, or be produced to meet in O, the rectangles $OQ\,.\,OQ'$, $Oq\,.\,Oq'$ are proportional to the squares on the parallel diameters.

But these rectangles are equal since QQ', qq' are chords of a circle.

Therefore the parallel diameters are equal, and, since equal diameters are equally inclined to the axes, it follows that the chords QQ', qq' are equally inclined to the axes.

Conversely, if two chords, not parallel, be equally inclined to the axes a circle can be drawn through their extremities.

For, as in Art. 92, if OQQ', Oqq' be two chords, and CD, Cd the parallel semi-diameters,

$$OQ\,.\,OQ' : Oq\,.\,Oq' :: CD^2 : Cd^2;$$

but, if CD and Cd be equally inclined to the axes, they are equal, and

$$\therefore OQ \,.\, OQ' = Oq \,.\, Oq',$$

and the points Q, Q', q, q' are concyclic.

EXAMPLES.

1. If the tangent at B meet the latus rectum produced in D, CDX is a right angle.

2. If PCp be a diameter, and the focal distance pS produced meet the tangent at P in T, $SP = ST$.

3. If the normal at P meet the axis minor in G' and $G'N$ be the perpendicular from G' on SP, then $PN = AC$.

4. The tangent at P bisects any straight line perpendicular to AA' and terminated by AP, $A'P$, produced if necessary.

5. Draw a tangent to an ellipse parallel to a given line.

6. SR being the semi-latus rectum, if RA meet the directrix in E, and $S'E$ meet the tangent at A in T,

$$AT = AS.$$

7. Prove that $SY : SP :: SR : PG$.

Find where the angle SPS' is greatest.

8. If two points E and E' be taken in the normal PG such that $PE = PE' = CD$, the loci of E and E' are circles.

9. If from the focus S' a line be drawn parallel to SP, it will meet the perpendicular SY in the circumference of a circle.

10. If the normal at P meet the axis major in G, prove that PG is an harmonic mean between the perpendiculars from the foci on the tangent at P.

11. The straight line NQ is drawn parallel to AP to meet CP in Q; prove that AQ is parallel to the tangent at P.

12. The locus of the intersection with the ordinate of the perpendicular from the centre on the tangent is an ellipse.

13. If a rectangle circumscribes an ellipse, its diagonals are the directions of conjugate diameters.

14. If tangents TP, TQ be drawn at the extremities, P, Q of any focal chord of an ellipse, prove that the angle PTQ is half the supplement of the angle which PQ subtends at the other focus.

15. If Y, Z be the feet of the perpendiculars from the foci on the tangent at P; prove that Y, N, Z, C are concyclic.

16. If AQ be drawn from one of the vertices perpendicular to the tangent at any point P, prove that the locus of the point of intersection of PS and QA produced will be a circle.

17. The straight lines joining each focus to the foot of the perpendicular from the other focus on the tangent at any point meet on the normal at the point and bisect it.

18. If two circles touch each other internally, the locus of the centres of circles touching both is an ellipse whose foci are the centres of the given circles.

19. The subnormal at any point P is a third proportional to the intercept of the tangent at P on the major axis and half the minor axis.

20. If the normal at P meet the axis major in G and the axis minor in g, $Gg : Sg :: SA : AX$, and if the tangent meet the axis minor in t,

$$St : tg :: BC : CD.$$

21. If the normal at a point P meet the axis in G, and the tangent at P meet the axis in T, prove that

$$TQ : TP :: BC : PG,$$

Q being the point where the ordinate at P meets the auxiliary circle.

22. If the tangent at any point P meet the tangent at the extremities of the axis AA' in F and F', prove that the rectangle AF, $A'F'$ is equal to the square on the semi-axis minor.

23. TP, TQ are tangents; prove that a circle can be described with T as centre so as to touch SP, HP, SQ, and HQ, or these lines produced, S and H being the foci.

24. If two equal and similar ellipses have the same centre, their points of intersection are at the extremities of diameters at right angles to one another.

25. The external angle between any two tangents to an ellipse is equal to the semi-sum of the angles which the chord joining the points of contact subtends at the foci.

26. The tangent at any point P meets the axes in T and t; if S be a focus the angles PSt, STP are equal.

27. A conic is drawn touching an ellipse at the extremities A, B of the axes, and passing through the centre C of the ellipse; prove that the tangent at C is parallel to AB.

28. The tangent at any point P is cut by any two conjugate diameters in T, t, and the points T, t are joined with the foci S, H respectively; prove that the triangles SPT, HPt are similar to each other.

29. If the diameter conjugate to CP meet SP, and HP (or these produced) in E and E', prove that SE is equal to HE', and that the circles which circumscribe the triangles SCE, HCE', are equal to one another.

30. PG is a normal, terminating in the major axis; the circle, of which PG is a diameter, cuts SP, HP, in K, L, respectively: prove that KL is bisected by PG, and is perpendicular to it.

31. Tangents are drawn from any point in a circle through the foci, prove that the lines bisecting the angles between the several pairs of tangents all pass through a fixed point.

32. If a quadrilateral circumscribe an ellipse, the angles subtended by opposite sides at one of the foci are together equal to two right angles.

33. If the normal at P meet the axis minor in G, and if the tangent at P meet the tangent at the vertex A in V, shew that

$$SG : SC :: PV : VA.$$

34. P, Q are points in two confocal ellipses, at which the line joining the common foci subtends equal angles; prove that the tangents at P, Q are inclined at an angle which is equal to the angle subtended by PQ at either focus.

35. The transverse axis is the greatest and the conjugate axis the least of all the diameters.

36. Prove that the locus of the centre of the circle inscribed in the triangle SPS' is an ellipse.

37. If the tangent and ordinate at P meet the transverse axis in T and N, prove that any circle passing through N and T will cut the auxiliary circle orthogonally.

38. If SY, $S'Y'$ be the perpendiculars from the foci on the tangent at a point P, and PN the ordinate, prove that

$$PY : PY' :: NY : NY'.$$

39. If a circle, passing through Y and Z, touch the major axis in Q, and that diameter of the circle, which passes through Q, meet the tangent in P, then $PQ = BC$.

40. From the centre of two concentric circles a straight line is drawn to cut them in P and Q; from P and Q straight lines are drawn parallel to two given lines at right angles. Shew that the locus of their point of intersection is an ellipse.

41. From any two points P, Q on an ellipse four lines are drawn to the foci S, S': prove that $SP \,.\, S'Q$ and $SQ \,.\, S'P$ are to one another as the squares of the perpendiculars from a focus on the tangents at P and Q.

42. Two conjugate diameters are cut by the tangent at any point P in M, N; prove that the area of the triangle CPM varies inversely as that of the triangle CPN.

43. If P be any point on the curve, and AV be drawn parallel to PC to meet the conjugate CD in V, prove that the areas of the triangles CAV, CPN are equal, PN being the ordinate.

44. Two tangents to an ellipse intersect at right angles; prove that the sum of the squares on the chords intercepted on them by the auxiliary circle is constant.

45. Prove that the distance between the two points on the circumference, at which a given chord, not passing through the centre, subtends the greatest and least angles, is equal to the diameter which bisects that chord.

46. The tangent at P intersects a fixed tangent in T; if S is the focus and a line be drawn through S perpendicular to ST, meeting the tangent at P in Q, shew that the locus of Q is a straight line touching the ellipse.

47. Shew that, if the distance between the foci be greater than the length of the axis minor, there will be four positions of the tangent, for which the area of the triangle, included between it and the straight lines drawn from the centre of the curve to the feet of the perpendiculars from the foci on the tangent, will be the greatest possible.

48. Two ellipses whose axes are equal, each to each, are placed in the same plane with their centres coincident, and axes inclined to each other. Draw their common tangents.

49. An ellipse is inscribed in a triangle, having one focus at the orthocentre; prove that the centre of the ellipse is the centre of the nine-point circle of the triangle and that its transverse axis is equal to the radius of that circle.

50. The tangent at any point P of a circle meets the tangent at a fixed point A in T, and T is joined with B the extremity of the diameter passing through A; the locus of the point of intersection of AP, BT is an ellipse.

51. The ordinate NP at a point P meets, when produced, the circle on the major axis in Q. If S be a focus of the ellipse, prove that $SQ : SP ::$ the axis major : the chord of the circle through Q and S, and that the diameter of the ellipse parallel to SP is equal to the same chord.

52. If the perpendicular from the centre C on the tangent at P meet the focal distance SP produced in R, the locus of R is a circle, the diameter of which is equal to the axis major.

53. A perfectly elastic billiard ball lies on an elliptical billiard table, and is projected in any direction along the table: shew that all the lines in which it moves after each successive impact touch an ellipse or an hyperbola confocal with the billiard table.

54. Shew that a circle can be drawn through the foci and the intersections of any tangent with the tangents at the vertices.

55. If CP, CD be conjugate semi-diameters, and a rectangle be described so as to have PD for a diagonal and its sides parallel to the axes, the other angular points will be situated on two fixed straight lines passing through the centre C.

56. If the tangent at P meet the minor axis in T, prove that the areas of the triangles SPS', STS' are in the ratio of the squares on CD and ST.

57. Find the locus of the centre of the circle touching the transverse axis, SP, and $S'P$ produced.

58. In an ellipse SQ and $S'Q$, drawn perpendicularly to a pair of conjugate diameters, intersect in Q; prove that the locus of Q is a concentric ellipse.

59. If the ordinate NP meet the auxiliary circle in Q, the perpendicular from S on the tangent at Q is equal to SP.

60. If PT, QT be tangents at corresponding points of an ellipse and its auxiliary circle, shew that

$$PT : QT :: BC : PF.$$

61. If CQ be conjugate to the normal at P, then is CP conjugate to the normal at Q.

62. PQ is one side of a parallelogram described about an ellipse, having its sides parallel to conjugate diameters, and the lines joining P, Q to the foci intersect in D, E; prove that the points D, E and the foci are concyclic.

63. If the centre, a tangent, and the transverse axis be given, prove that the directrices pass each through a fixed point.

64. The straight line joining the feet of perpendiculars from the focus on two tangents is at right angles to the line joining the intersection of the tangents with the other focus.

65. A circle passes through a focus, has its centre on the major axis of the ellipse, and touches the ellipse: shew that the straight line from the focus to the point of contact is equal to the latus rectum.

66. Prove that the perimeter of the quadrilateral formed by the tangent, the perpendiculars from the foci, and the transverse axis, will be the greatest possible when the focal distances of the point of contact are at right angles to each other.

67. Given a focus, the length of the transverse axis, and that the second focus lies on a straight line, prove that the ellipse will touch two fixed parabolas having the given focus for focus.

68. Tangents are drawn from a point on one of the equi-conjugate diameters; prove that the point, the centre, and the two points of contact are concyclic.

69. If PN be the ordinate of P, and if with centre C and radius equal to PN a circle be described intersecting PN in Q, prove that the locus of Q is an ellipse.

70. If AQO be drawn parallel to CP, meeting the curve in Q and the minor axis in O, $2CP^2 = AO \,.\, AQ$.

71. PS is a focal distance; CR is a radius of the auxiliary circle parallel to PS, and drawn in the direction from P to S; SQ is a perpendicular on CR: shew that the rectangle contained by SP and QR is equal to the square on half the minor axis.

72. If a focus be joined with the point where the tangent at the nearer vertex intersects any other tangent, and perpendiculars be let fall from the other focus on the joining line and on the last-mentioned tangent, prove that the distance between the feet of these perpendiculars is equal to the distance from either focus to the remoter vertex.

73. A parallelogram is described about an ellipse; if two of its angular points lie on the directrices, the other two will lie on the auxiliary circle.

74. From a point in the auxiliary circle straight lines are drawn touching the ellipse in P and P'; prove that SP is parallel to $S'P'$.

75. Find the locus of the points of contact of tangents to a series of confocal ellipses from a fixed point in the axis major.

76. A series of confocal ellipses intersect a given straight line; prove that the locus of the points of intersection of the pairs of tangents drawn at the extremities of the chords of intersection is a straight line at right angles to the given straight line.

77. Given a focus and the length of the major axis; describe an ellipse touching a given straight line and passing through a given point.

78. Given a focus and the length of the major axis; describe an ellipse touching two given straight lines.

79. Find the positions of the foci and directrices of an ellipse which touches at two given points P, Q, two given straight lines PO, QO, and has one focus on the line PQ, the angle POQ being less than a right angle.

80. Through any point P of an ellipse are drawn straight lines APQ, $A'PR$, meeting the auxiliary circle in Q, R, and ordinates Qq, Rr are drawn to the transverse axis; prove that, L being an extremity of the latus rectum,

$$Aq \,.\, A'r : Ar \,.\, A'q :: AC^2 : SL^2.$$

81. If a tangent at a point P meet the major axis in T, and the perpendiculars from the focus and centre in Y and Z, then

$$TY^2 : PY^2 :: TZ : PZ.$$

82. An ellipse slides between two lines at right angles to each other; find the locus of its centre.

83. TP, TQ are two tangents, and CP', CQ' are the radii from the centre respectively parallel to these tangents, prove that $P'Q'$ is parallel to PQ.

84. The tangent at P meets the minor axis in t; prove that

$$St \,.\, PN = BC \,.\, CD.$$

85. If the circle, centre t, and radius tS, meet the ellipse in Q, and QM be the ordinate, prove that

$$QM : PN :: BC : BC + CD.$$

86. Perpendiculars SY, $S'Y'$ are let fall from the foci upon a pair of tangents TY, TY'; prove that the angles STY, $S'TY'$ are equal to the angles at the base of the triangle YCY'.

87. PQ is the chord of an ellipse normal at P, LCL' the diameter bisecting it, shew that PQ bisects the angle LPL' and that $LP + PL'$ is constant.

88. ABC is an isosceles triangle of which the side AB is equal to the side AC. BD, BE drawn on opposite sides of BC and equally inclined to it meet AC in D and E. If an ellipse is described round BDE having its axis minor parallel to BC, then AB will be a tangent to the ellipse.

89. If A be the extremity of the major axis and P any point on the curve, the bisectors of the angles PSA, $PS'A$ meet on the tangent at P.

90. If two ellipses intersect in four points, the diameters parallel to a pair of the chords of intersection are in the same ratio to each other.

91. From any point P of an ellipse a straight line PQ is drawn perpendicular to the focal distance SP, and meeting in Q the diameter conjugate to that through P; shew that PQ varies inversely as the ordinate of P.

92. If a tangent to an ellipse intersect at right angles a tangent to a confocal ellipse, the point of intersection lies on a fixed circle.

93. If from a point T in the director circle of an ellipse tangents TP, TP' are drawn, the line joining T with the intersection of the normals at P and P' passes through C.

94. Through the middle point of a focal chord a straight line is drawn at right angles to it to meet the axis in R; prove that SR bears to SC the duplicate ratio of the chord to the diameter parallel to it, S being the focus and C the centre.

95. The tangent at a point P meets the auxiliary circle in Q' to which corresponds Q on the ellipse; prove that the tangent at Q cuts the auxiliary circle in the point corresponding to P.

96. If a chord be drawn to a series of concentric, similar, and similarly situated ellipses, and meet one in P and Q, and if on PQ as diameter a circle be described meeting that ellipse again in RS, shew that RS is constant in position for all the ellipses.

97. An ellipse touches the sides of a triangle; prove that if one of its foci move along the arc of a circle passing through two of the angular points of the triangle, the other will move along the arc of a circle through the same two angular points.

98. The normal at a point P of an ellipse meets the conjugate axis in K, and a circle is described with centre K and passing through the foci S and H. The lines SQ, HQ, drawn through any point Q of this circle, meet the tangent at P in T and t; prove that T and t lie on a pair of conjugate diameters.

99. If SP, $S'Q$ be parallel focal distances drawn towards the same parts, the tangents at P and Q intersect on the auxiliary circle.

100. Having given one focus, one tangent and the eccentricity of an ellipse, prove that the locus of the other focus is a circle.

101. PSQ is a focal chord of an ellipse, and pq is any parallel chord; if PQ meet in T the tangent at p,

$$pq : PQ :: Sp : ST.$$

102. If an ellipse be inscribed in a quadrilateral so that one focus is equidistant from the four vertices, the other focus must be at the intersection of the diagonals.

103. If a pair of conjugate diameters of an ellipse be produced to meet either directrix, prove that the orthocentre of the triangle so formed is the corresponding focus of the curve.

104. A pair of conjugate diameters intercept, on the tangent at either vertex, a length which subtends supplementary angles at the foci.

105. The straight lines TP, TQ are the tangents at the points P, Q of an ellipse; one circle touches TP at P and meets TQ in Q and Q', and another circle touches TQ at Q and meets TP in P and P'; prove that PQ' and $P'Q$ are parallel, and that they are divided in the same ratio by the ellipse.

106. If the normals at P and D meet in E, prove that EC is perpendicular to PD, and that the straight line joining C to the centroid of the triangle EPD bisects the line joining E to T, the point of intersection of the tangents at P and D.

107. A chord PQ, normal at P, meets the directrices in K and L, and the tangents at P and Q meet in T; prove that PK and QL subtend equal angles at T, and that KL subtends at T an angle which is half the sum of the angles subtended by SS' at the ends of the chord.

108. The tangent at the point P meets the directrices in E and F; prove that the other tangents from E and F intersect on the normal at P.

109. If the tangent at any point meets a pair of conjugate diameters in T and T', prove that TT' subtends supplementary angles at the foci.

110. PSQ, $PS'R$ are focal chords; prove that the tangent at P and the chord QR cut the major axis at equal distances from the centre.

CHAPTER IV.

THE HYPERBOLA.

DEFINITION.

An hyperbola is the curve traced by a point which moves in such a manner, that its distance from a given point is in a constant ratio of greater inequality to its distance from a given straight line.

Tracing the Curve.

94. Let S be the focus, EX the directrix, and A the vertex.

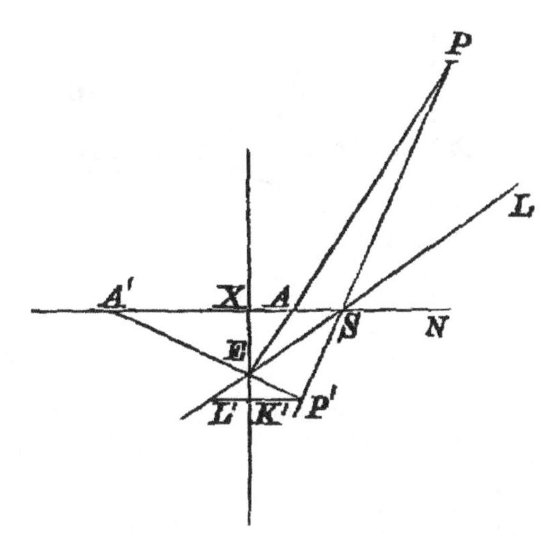

Then, as in Art. 1, any number of points on the curve may be obtained by taking successive positions of E on the directrix.

In SX produced, find a point A' such that

$$SA' : A'X :: SA : AX,$$

then A' is the other vertex as in the ellipse, and, the eccentricity being greater than unity, the points A and A' are evidently on opposite sides of the directrix.

Find the point P corresponding to E, and let $A'E$, PS produced meet in P', then, if $P'K'$ perpendicular to the directrix meet SE produced in L',

$$P'L' : P'K' :: SA' : A'X :: SA : AX,$$

and the angle

$$P'L'S = L'SX = L'SP';$$

$$\therefore SP' = P'L'.$$

Hence P' is a point in the curve, and PSP' is a focal chord.

Following out the construction we observe that, since SA is greater than AX, there are two points on the directrix, e and e', such that Ae and Ae' are each equal to AS.

If E coincide with e, the angle

$$QSL = LSN = ASe = AeS.$$

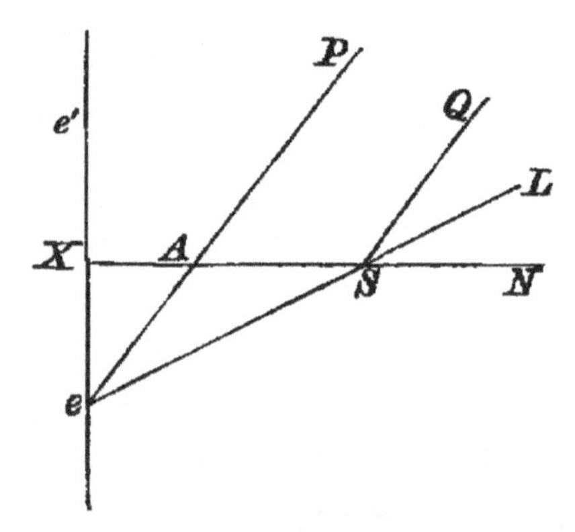

Hence SQ, AP are parallel, and the corresponding point of the curve is at an infinite distance; and similarly the curve tends to infinity in the direction Ae'.

Further, the angle ASE is less or greater than AES, according as the point E is, or is not, between e and e'.

Hence, when E is below e, the curve lies above the axis, to the right of the directrix; when between e and X, below the axis to the left; when between

X and e', above the axis to the left; and when above e', below the axis to the right. Hence a general idea can be obtained of the form of the curve, tending to infinity in four directions, as in the figure of Art. 102.

DEFINITIONS.

The line AA' is called the transverse axis of the hyperbola.

The middle point, C, of AA' is the centre.

Any straight line, drawn through C and terminated by the curve, is called a diameter.

95. Prop. I. *If P be any point of an hyperbola, and AA' its transverse axis, and if $A'P$, and PA produced, (or PA and PA' produced) meet the directrix in E and F, EF subtends a right angle at the focus.*

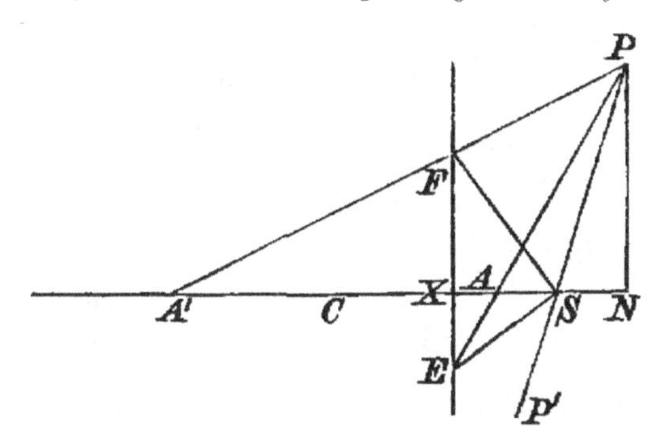

By the theorem of Art. 4, ES' bisects the angle ASP' and FS bisects ASP;

$$\therefore ESF \text{ is a right angle.}$$

SAA' being a focal chord, this is a particular case of the theorem of Art. 6.

96. Prop. II. *If PN be the ordinate of a point P, and ACA' the transverse axis, PN^2 is to $AN \,.\, NA'$ in a constant ratio.*

Join AP, $A'P$, meeting the directrix in E and F.

Then $$PN : AN :: EX : AX,$$

and $$PN : A'N :: FX : A'X;$$

$$\therefore PN^2 : AN \,.\, NA' :: EX \,.\, FX : AX \,.\, A'X$$
$$:: SX^2 : AX \,.\, A'X,$$

since ESF is a right angle; that is, PN^2 is to $AN \,.\, NA'$, in a constant ratio.

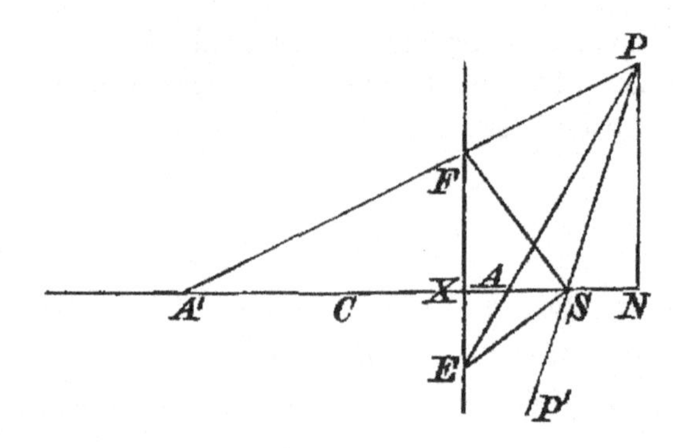

Through C, the middle point of AA', draw CB at right angles to the axis, and such that

$$BC^2 : AC^2 :: SX^2 : AX \,.\, A'X;$$

then
$$PN^2 : AN \,.\, NA' :: BC^2 : AC^2$$

or
$$PN^2 : CN^2 - AC^2 :: BC^2 - AC^2$$

COR. If PM be the perpendicular from P to BC

$$PM = CN, \text{ and } PN = CM;$$

$$\therefore CM^2 : PM^2 - AC^2 :: BC^2 : AC^2$$

or
$$CM^2 : BC^2 :: PM^2 - AC^2 : AC^2$$

$$\therefore CM^2 + BC^2 : BC^2 :: PM^2 : AC^2$$

or
$$PM^2 : CM^2 + BC^2 :: AC^2 : BC^2$$

97. If we describe the circle on AA' as diameter, which we may term, for convenience, *the auxiliary circle*, the rectangle $AN \,.\, NA'$ is equal to the square on the tangent to the circle from N.

Hence the preceding theorem may be thus expressed:

The ordinate of an hyperbola is to the tangent from its foot to the auxiliary circle in the ratio of the conjugate to the transverse axis.

DEF. *If CB' be taken equal to CB, on the other side of the axis, the line BCB' is called the conjugate axis.*

The two lines AA', BB' are the principal axes of the curve.

When these lines are equal, the hyperbola is said to be equilateral, or rectangular.

The lines AA', BB' are sometimes called major and minor axes, but, as AA' is not necessarily greater than BB', these terms cannot with propriety be generally employed.

If a point N' be taken on CA' produced, such that $CN' = CN$, the corresponding ordinate $P'N' = PN$, and therefore it follows that the curve is symmetrical with regard to BCB', and that there is another focus and directrix, corresponding to the vertex A'.

98. Prop. III. *If ACA' be the transverse axis, C the centre, S one of the foci, and X the foot of the directrix,*

$$CS : CA :: CA : CX :: SA : AX,$$

and

$$CS : CX :: CS^2 : CA^2.$$

Interchanging the positions of S and X for a new

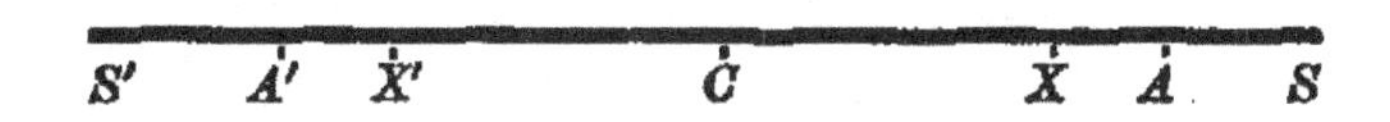

figure, the proof of these relations is identical with the proof given for the ellipse in Art. 61.

99. Prop. IV. *If S be a focus, and B an extremity of the conjugate axis,*

$$BC^2 = AS \,.\, SA', \text{ and } SC^2 = AC^2 + BC^2.$$

Referring to Art. (98), $SX = SA + AX$;

$$\therefore SX : AX :: SA + AX : AX,$$
$$:: SC + AC : AC;$$

and similarly

$$SX : A'X :: SC - AC : AC;$$
$$\therefore SX^2 : AX \,.\, A'X :: SC^2 - AC^2 : AC^2.$$

But

$$BC^2 : AC^2 :: SX^2 : AX \,.\, A'X;$$
$$\therefore BC^2 = SC^2 - AC^2 = AS \,.\, SA'.$$

Hence

$$SC^2 = AC^2 + BC^2 = AB^2;$$

i.e. SC is equal to the line joining the ends of the axes.

100. Prop. V. *The difference of the focal distances of any point is equal to the transverse axis.*

For, if PKK', perpendicular to the directrices, meet them in K and K',

$$S'P : PK' :: SA : AX,$$

and

$$SP : PK :: SA : AX;$$

$$\therefore S'P - SP : KK' :: SA : AX,$$

$$:: AA' : XX' \text{ (Art. 98)};$$

$$\therefore S'P - SP = AA'.$$

Cor. 1.

$$SP : NX :: AC : CX;$$

$$\therefore SP : AC :: NX : CX;$$

$$\therefore SP + AC : AC :: CN : CX,$$

or

$$SP + AC : CN :: SA : AX.$$

Hence also

$$S'P - AC : CN :: SA : AX.$$

Cor. 2. *Hence also it can be easily shewn, that the difference of the distances of any point from the foci of an hyperbola, is greater or less than the transverse axis, according as the point is within or without the concave side of the curve.*

101. *Mechanical Construction of the Hyperbola.*

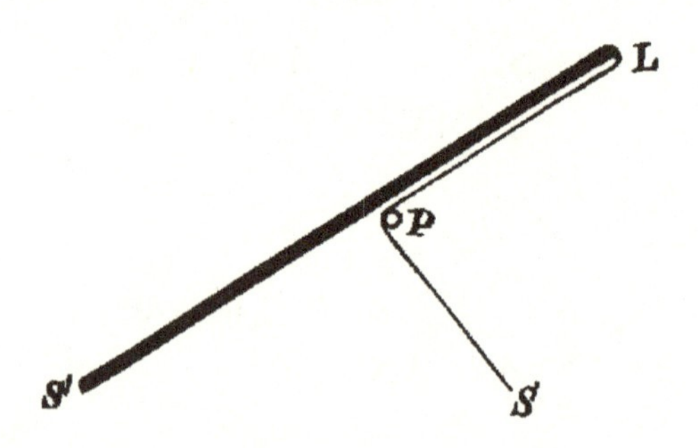

Let a straight rod $S'L$ be moveable in the plane of the paper about the point S'. Take a piece of string, the length of which is less than that of the rod, and fasten one end to a fixed point S, and the other end to L; then, pressing a pencil against the string so as to keep it stretched, and a part of it PL in contact with the rod, the pencil will trace out on the paper an hyperbola, having its foci at S and S', and its transverse axis equal to the difference between the length of the rod and that of the string.

This construction gives the right-hand branch of the curve; to trace the other branch, take the string longer than the rod, and such that it exceeds the length of the rod by the transverse axis.

We may remark that by taking a longer rod $MS'L$, and taking the string longer than $SS' + S'L$, so that the point P will be always on the end $S'M$ of the rod, we shall obtain an ellipse of which S and S' are the foci. Moreover, remembering that a parabola is the limiting form of an ellipse when one of the foci is removed to an infinite distance, the mechanical construction given for the parabola will be seen to be a particular case of the above.

The Asymptotes.

102. We have shewn in Art. 94 that if two points, e and e', be taken on the directrix such that

$$Ae = Ae' = AS,$$

the lines eA, $e'A$ meet the curve at an infinite distance.

These lines are parallel to the diagonals of the rectangle formed by the axes, for

$$Ae' : AX :: AS : AX :: SC : AC,$$
$$:: AB : AC, \text{ (Art. 99)}.$$

DEFINITION. *The diagonals of the rectangle formed by the principal axes are called the asymptotes.*

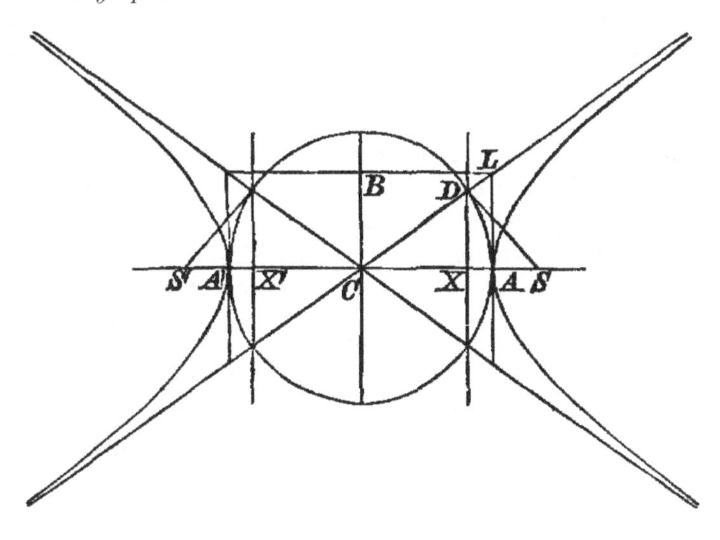

We observe that the axes bisect the angles between the asymptotes, and that if a double ordinate, PNP', when produced, meet the asymptotes in Q and Q',

$$PQ = P'Q'.$$

The figure appended will give the general form of the curve and its connection with the asymptotes and the auxiliary circle.

103. PROP. VI. *The asymptotes intersect the directrices in the same points as the auxiliary circle, and the lines joining the corresponding foci with the points of intersection are tangents to the circle.*

If the asymptote CL meet the directrix in D, joining SD (fig. Art. 102), $CL^2 = AC^2 + BC^2 = SC^2$,

and $$CD : CX :: CL : CA :: SC : CA :: CA : CX;$$

$\therefore CD = CA$, and D is on the auxiliary circle.

Also

$$CS \,.\, CX = CA^2 = CD^2;$$

$\therefore CDS$ is a right angle, and SD is the tangent at D.

COR. $CD^2 + SD^2 = CS^2 = AC^2 + BC^2$ (Art. 99);

$$\therefore SD = BC.$$

104. An asymptote may also be characterized as the ultimate position of a tangent when the point of contact is removed to an infinite distance.

It appears from Art. 10 that in order to find the point of contact of a tangent drawn from a point T in the directrix, we must join T with the focus S, and draw through S a straight line at right angles to ST; this line will meet the curve in the point of contact.

In the figures of Arts. 94 and 102 we know that the line through S, parallel to eA or CL, meets the curve in a point at an infinite distance, and also that this straight line is at right angles to SD, since SD is at right angles to CD. Hence the tangent from D, that is the line from D to the point at an infinite distance, is perpendicular to DS and therefore coincident with CD.

The asymptotes therefore touch the curve at an infinite distance.

105. DEF. *If an hyperbola be described, having for its transverse and conjugate axes, respectively, the conjugate and transverse axes of a given hyperbola, it is called the conjugate hyperbola.*

It is evident from the preceding article that the conjugate hyperbola has the same asymptotes as the original hyperbola, and that the distances of its foci from the centre are also the same.

The relations of Art. 96 and its Corollary are also true, *mutatis mutandis*, of the conjugate hyperbola; thus, if R be a point in the conjugate hyperbola,

$$RM^2 : CM^2 - BC^2 :: AC^2 : BC^2,$$

and

$$CM^2 : RM^2 + AC^2 :: BC^2 : AC^2.$$

DEF. *A straight line drawn through the centre and terminated by the conjugate hyperbola is also called a diameter of the original hyperbola.*

106. PROP. VII. *If from any point Q in one of the asymptotes, two straight lines QPN, QRM be drawn at right angles respectively to the transverse and conjugate axes, and meeting the hyperbola in P, p, and the conjugate hyperbola in R, r,*

$$QP \,.\, Qp = BC^2,$$

and

$$QR \,.\, Qr = AC^2.$$

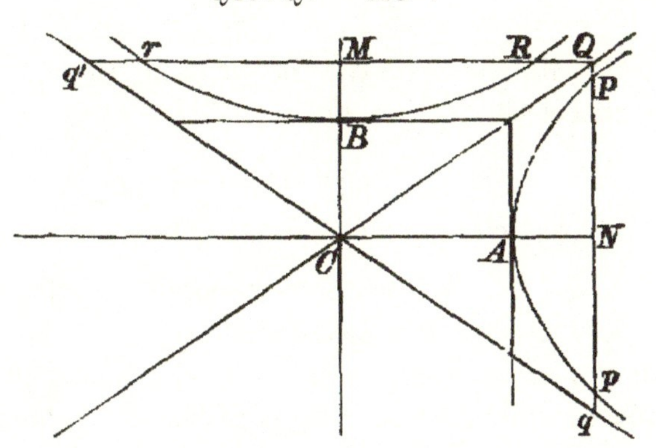

For

$$QN^2 : BC^2 :: CN^2 : AC^2;$$

$$\therefore QN^2 - BC^2 : BC^2 :: CN^2 - AC^2 : AC^2$$

$$:: PN^2 : BC^2;$$

$$\therefore QN^2 - BC^2 = PN^2,$$

or

$$QN^2 - PN^2 = BC^2;$$

i.e.

$$QP \,.\, Qp = BC^2.$$

Similarly,

$$QM^2 : AC^2 :: CM^2 : BC^2;$$

$$\therefore QM^2 - AC^2 : AC^2 :: CM^2 - BC^2 : BC^2,$$

$$:: RM^2 : AC^2;$$

$$\therefore QM^2 - RM^2 = AC^2,$$

or

$$QR \,.\, Qr = AC^2.$$

These relations may also be given in the form,

$$QP \,.\, Pq = BC^2, \quad QR \,.\, Rq' = AC^2.$$

Cor. If the point Q be taken at a greater distance from C, the length QN and therefore Qp will be increased, and may be increased indefinitely.

But the rectangle $QP \,.\, Qp$ is of finite magnitude; hence QP will be indefinitely diminished, and the curve, therefore, as it recedes from the centre, tends more and more nearly to coincide with the asymptote.

A further illustration is thus given of the remarks in Art. 104.

107. If in the preceding figure the line Qq be produced to meet the conjugate hyperbola in E and e, it can be shewn, in the same manner as in Art. 106, that

$$QE \,.\, Qe = BC^2;$$

and this equality is still true when the line Qq lies between C and A, in which case Qq does not meet the hyperbola.

Properties of the Tangent and Normal.

108. In the case of the hyperbola the theorem, proofs of which are given in Arts. 15 and 16, takes the following form:

The tangents drawn from any point to an hyperbola subtend equal or supplementary angles at either focus according as they touch the same or opposite branches of the curve.

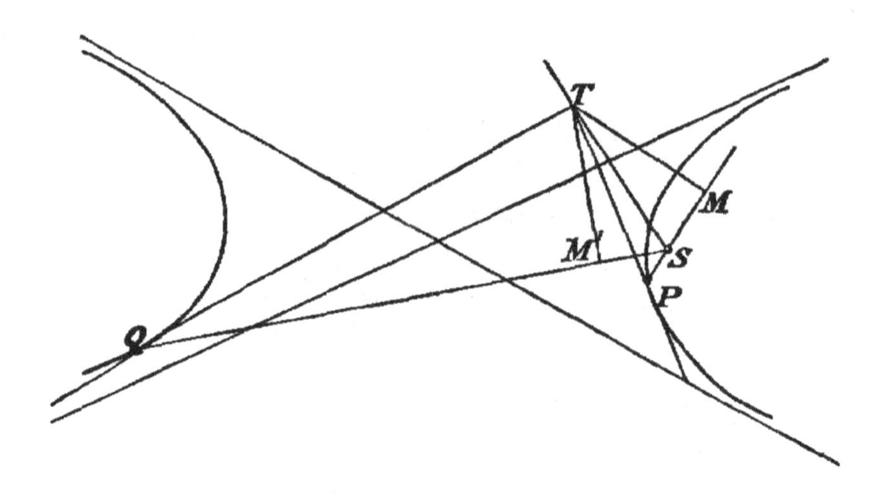

For, T being the point of intersection of tangents to opposite branches of the curve, let TM, TM' be the perpendiculars let fall from T on SP and SQ, then, as in Arts. 15 and 16, $TM = TM'$;

$\therefore$ the angles TSM, TSM' are equal, and consequently the angles TSP, TSQ are supplementary.

109. PROP. VIII. *The tangent at any point bisects the angle between the focal distances of that point, and the normal is equally inclined to the focal distances.*

Let the normal at P meet the axis in G.

Then (Art. 18),

$$SG : SP :: SA : AX,$$

and
$$S'G : S'P :: SA : AX;$$

$$\therefore SG : S'G :: SP : S'P;$$

and therefore the angle between SP and $S'P$ produced is bisected by PG.

Hence PT, the tangent which is perpendicular to PG, bisects the angle SPS'.

COR. 1. If PT and GP produced meet, respectively, the conjugate axis in t and g, it can be shewn, in exactly the same manner as in the corresponding case of the ellipse (Art. 68), that S, P, S', t, and g are concyclic.

COR. 2. If an ellipse be described having S and S' for its foci, and if this ellipse meet the hyperbola in P, the normal at P to the ellipse bisects the angle SPS', and therefore coincides with the tangent to the hyperbola.

Hence, if an ellipse and an hyperbola be confocal, that is, have the same foci, they intersect at right angles.

110. PROP. IX. *Every diameter is bisected at the centre, and the tangents at the ends of a diameter are parallel.*

Let PCp be a diameter, and PN, pn the ordinates.

Then
$$CN^2 : Cn^2 :: PN^2 : pn^2,$$
$$:: CN^2 - AC^2 : Cn^2 - AC^2;$$

hence $CN = Cn$, and $\therefore CP = Cp$.

Again, if PT, pt be the tangents,

The triangles PCS, pCS' are equal in all respects, and therefore $SPS'p$ is a parallelogram.

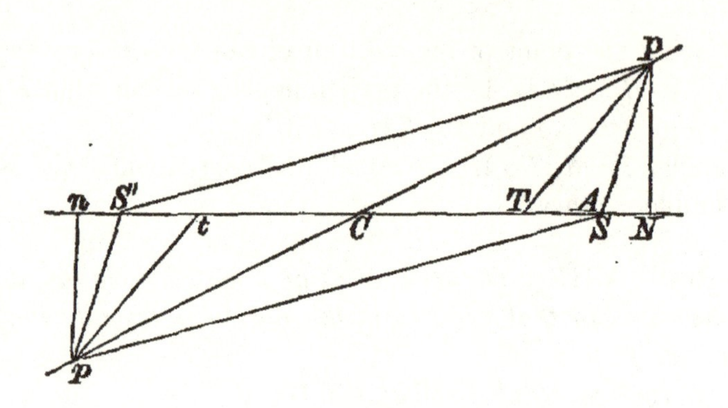

Hence the angles SPS', SpS' are equal, and therefore $SPT = S'pt$.

But $SPC = S'pC$,

$\therefore$ the difference TPC = the difference tpC, and PT is parallel to pt.

It can be shewn in exactly the same manner, that, if the diameter be terminated by the conjugate hyperbola, it is bisected in C, and the tangents at its extremities are parallel.

COR. The distances SP, Sp are equally inclined to the tangents at P and p.

111. PROP. X. *The perpendiculars from the foci on any tangent meet the tangent on the auxiliary circle, and the semi-conjugate axis is a mean proportional between their lengths.*

Let SY, $S'Y'$ be the perpendiculars, and let SY produced meet $S'P$ in L.

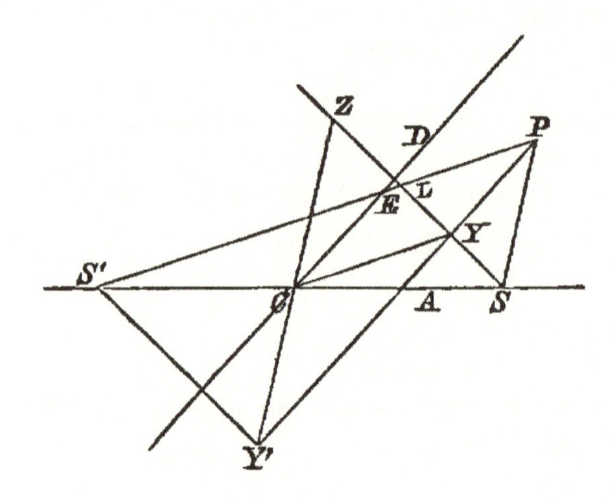

Then the triangles SPY, LPY are equal in all respects,

$$\text{and } SY = LY.$$

Hence, C being the middle point of SS' and Y of SL, CY is parallel to $S'L$, and $S'L = 2CY$.

But $$S'L = S'P - PL = S'P - SP = 2AC;$$

$$\therefore CY = AC,$$

and Y is on the auxiliary circle.

So also Y' is a point in the circle.

Let SY produced meet the circle in Z, and join $Y'Z$; then, $Y'YZ$ being a right angle, ZY' is a diameter and passes through C. Hence, the triangles SCZ, $S'CY'$ being equal,

$$S'Y' = SZ,$$

and $$SY \,.\, S'Y' = SY \,.\, SZ = SA \,.\, SA' = BC^2.$$

Cor. 1. If P' be the other extremity of the diameter PC, the tangent at P' is parallel to PY, and therefore Z is the foot of the perpendicular from S on the tangent at P'.

Cor. 2. If the diameter DCD', drawn parallel to the tangent at P, meet $S'P$, SP in E and E', $PECY$ is a parallelogram;

$$\therefore PE = CY = AC,$$

and so also $$PE' = CY' = AC.$$

112. Prop. XI. *To draw tangents to an hyperbola from a given point.*

The construction of Art. 17 may be employed, or, as in the cases of the ellipse and parabola, the following.

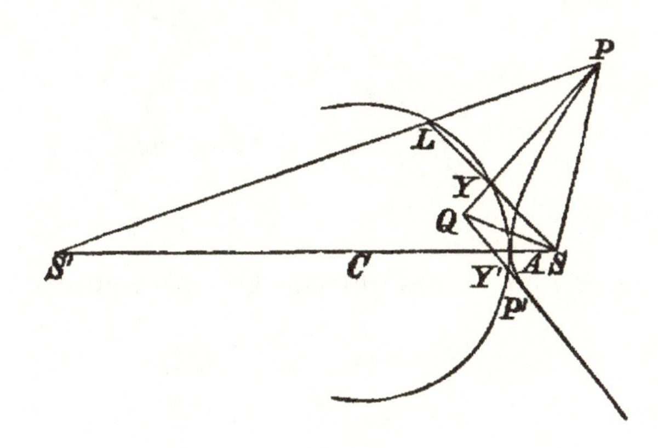

Let Q be the given point; join SQ, and upon SQ as diameter describe a circle intersecting the auxiliary circle in Y and Y';

$$QY \text{ and } QY' \text{ are the required tangents.}$$

Producing SY to L, so that $YL = SY$, draw $S'L$ cutting QY in P, and join SP.

The triangles SPY, LPY are equal in all respects,

and
$$S'P - SP = S'L = 2CY = 2AC;$$

$$\therefore P \text{ is a point on the hyperbola.}$$

Also QP bisects the angle SPS', and is therefore the tangent at P. A similar construction will give the other tangent QP'.

If the point Q be within the angle formed by the asymptotes, the tangents will both touch the same branch of the curve; but if it lie within the external angle, they will touch opposite branches.

113. PROP. XII. *If two tangents be drawn from any point to an hyperbola they are equally inclined to the focal distances of that point.*

Let PQ, $P'Q$ be the tangents, SY, $S'Y'$, SZ, $S'Z'$ the perpendiculars from the foci; join YZ, $Y'Z'$.

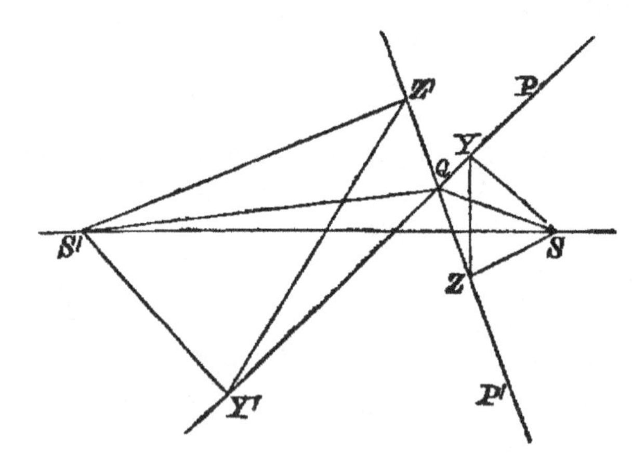

Then the angles YSZ, $Y'S'Z'$ are equal, for they are the supplements of YQZ, $Y'QZ'$.

Also
$$SY \,.\, S'Y' = SZ \,.\, S'Z' \text{ (Art. 111);}$$

or
$$SY : SZ :: S'Z' : S'Y';$$

$$\therefore \text{the triangles } YZS,\ Y'S'Z' \text{ are similar,}$$
$$\text{and the angle } YZS = Z'Y'S'.$$
$$\text{But the angle } YQS = YZS, \text{ and } Z'QS' = ZY'S';$$
$$\therefore YQS = Z'QS'.$$

That is, the tangent QP and the tangent $P'Q$ produced are equally inclined to SQ and $S'Q$.

Or, producing $S'Q$, QP and QP' are equally inclined to QS and $S'Q$ produced.

In exactly the same manner it can be shewn that if QP, QP' touch opposite branches of the curve the angles PQS, $P'QS'$ are equal.

Cor. If Q be a point in a confocal hyperbola, the normal at Q bisects the angle between SQ and $S'Q$ produced and therefore bisects the angle PQP'.

Hence, if from any point of an hyperbola tangents be drawn to a confocal hyperbola, these tangents are equally inclined to the normal or the tangent at the point, according as it lies within or without that angle formed by the asymptotes of the confocal which contains the transverse axes.

114. Prop. XIII. *If PT, the tangent at P, meet the transverse axis in T, and PN be the ordinate,*

$$CN\,.\,CT = AC^2$$

Let fall the perpendicular SY upon PT, and join YN, CY, SP, and $S'P$.

The angle $\quad CYT = S'PY = SPY$

$$= \text{the supplement of } SNY = CNY;$$

also the angle YCT is common to the two triangles CYT, CYN; these triangles are therefore similar, and

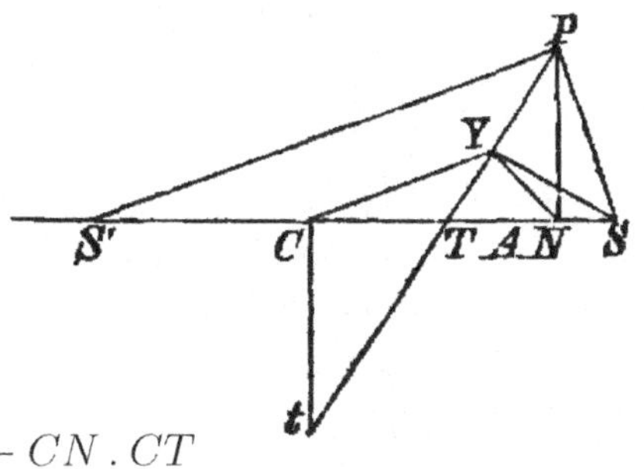

$$CN : CY :: CY : CT,$$

or

$$CN\,.\,CT = CY^2 = AC^2$$

Cor. 1. Hence $\quad CN\,.\,NT = CN^2 - CN\,.\,CT$

$$= CN^2 - AC^2$$
$$= AN\,.\,NA'$$

COR. 2. Hence also it follows that

If any number of hyperbolas be described having the same transverse axis, and an ordinate be drawn cutting the hyperbolas, the tangents at the points of section will all meet the transverse axis in the same point.

COR. 3. If CN be increased indefinitely, CT is diminished indefinitely, and the tangent ultimately passes through C, as we have already shewn in Art. 104.

115. PROP. XIV. *If the tangent at P meet the conjugate axis in t, and PN be the ordinate,*

$$Ct \,.\, PN = BC^2.$$

For $$Ct : PN :: CT : NT; \text{ (Fig. Art. 114)}$$

$$\therefore Ct \,.\, PN : PN^2 :: CT \,.\, CN : CN \,.\, NT$$
$$:: AC^2 : AN \,.\, NA'.$$
$$\therefore Ct \,.\, PN : AC^2 :: PN^2 : AN \,.\, NA'$$
$$:: BC^2 : AC^2,$$

and $$Ct \,.\, PN = BC^2.$$

In exactly the same manner as in Art. 76, it can be shewn that

$$CG \,.\, CT = SC^2,$$

$$CG : CN :: SC^2 : AC^2, \quad Cg : PN :: SC^2 : BC^2,$$

and $$NG : CN :: BC^2 : AC^2.$$

116. PROP. XV. *If the normal at P meet the transverse axis in G, the conjugate axis in g, and the diameter parallel to the tangent at P in F,*

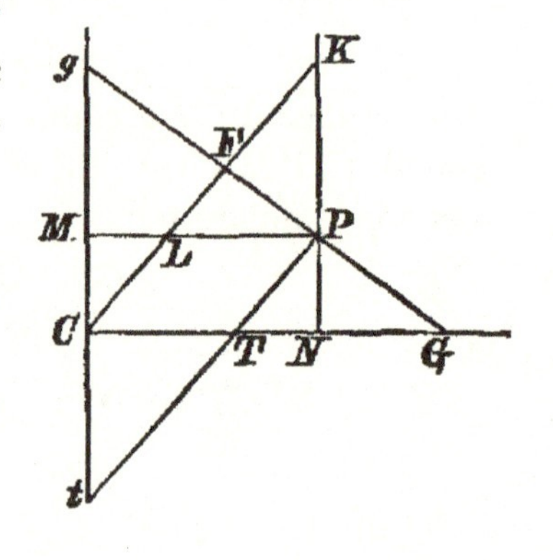

$$PF \,.\, PG = BC^2, \text{ and } PF \,.\, Pg = AC^2.$$

Let NP, PM, perpendicular to the axes, meet the diameter CF in K and L;

Then KNG, KFG being right angles, K, F, N, G are concyclic;

$$\therefore PF \,.\, PG = PK \,.\, PN$$
$$= Ct \,.\, PN = BC^2.$$

Similarly F, L, M, g are concyclic;

$$\therefore PF \,.\, Pg = PL \,.\, PM = CT \,.\, CN = AC^2.$$

117. PROP. XVI. *If PCp be a diameter, and QV an ordinate, and if the tangent at Q meet the diameter Pp in T,*

$$CV \, . \, CT = CP^2.$$

Let the tangents at P and p meet the tangent at Q in R and r;

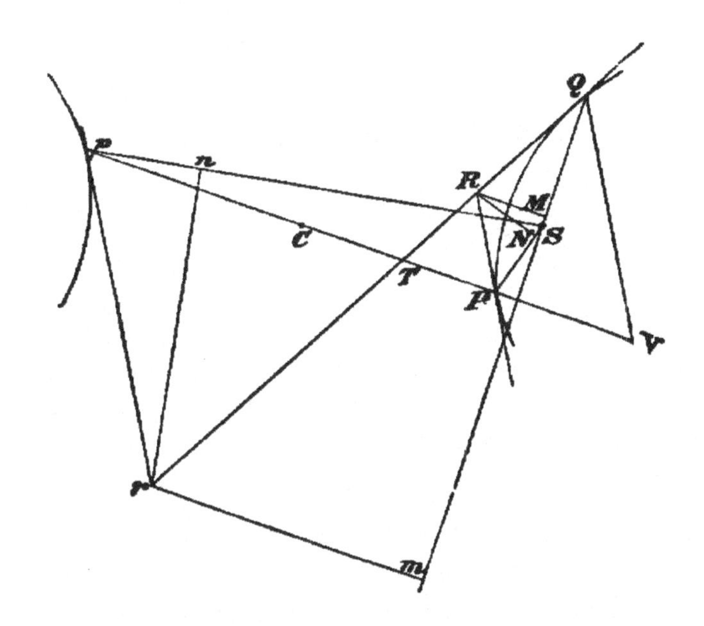

Then the angle $SPR = Spr$ (Cor. Art. 110)
and therefore if RN, rn be the perpendiculars on SP, sp, the triangles RPN, rpn are similar.

Draw RM, rm perpendiculars on SQ.

Then $\quad TR : Tr :: RP : rp :: RN : rn,$

$$:: RM : rm \text{ (Cor. Art. 15)}$$

$$:: RQ : rQ.$$

Hence, QV, RP, and rp being parallel,

$$TP : Tp :: PV : pV;$$

$$\therefore TP + Tp : Tp - TP :: PV + pV : pV - PV,$$

or
$$2CP : 2CT :: 2CV : 2CP,$$

or
$$CV \, . \, CT = CP^2.$$

118. Prop. XVII. *A diameter bisects all chords parallel to the tangents at its extremities.*

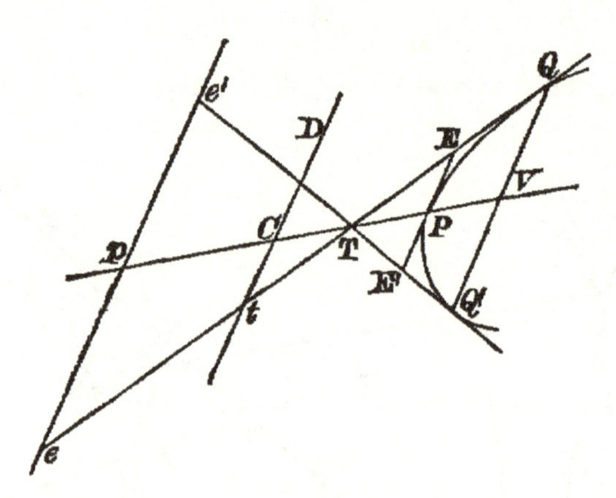

Let PCp be the diameter, and QQ' the chord, parallel to the tangents at P and p. Then if the tangents TQ, TQ' at Q and Q' meet the tangents at P and p, in the points E, E', e, e',

$$EP = E'P \text{ and } ep = e'p, \text{ (Art. 21)}$$

$$\therefore \text{the point } T \text{ is on the line } Pp;$$

$$\text{but } TP \text{ bisects } QQ';$$

that is, the diameter pCP produced bisects QQ'.

Def. *The line DCd, drawn parallel to the tangent at P and terminated by the conjugate hyperbola, that is, the diameter parallel to the tangent at P, is said to be conjugate to PCp.*

A diameter therefore bisects all chords parallel to its conjugate.

119. Prop. XVIII. *If the diameter DCd be conjugate to PCp, then will PCp be conjugate to DCd.*

Let the chord QVq be parallel to CD and be bisected in V by CP produced.

Draw the diameter qCR, and join RQ meeting CD in U.

Then $RC = Cq$ and $QV = Vq$; $\therefore QR$ is parallel to CP.

Also $$QU : UR :: Cq : CR,$$

and $$\therefore QU = UR,$$

that is, CD bisects the chords parallel to CP, and PCp is therefore conjugate to DCd.

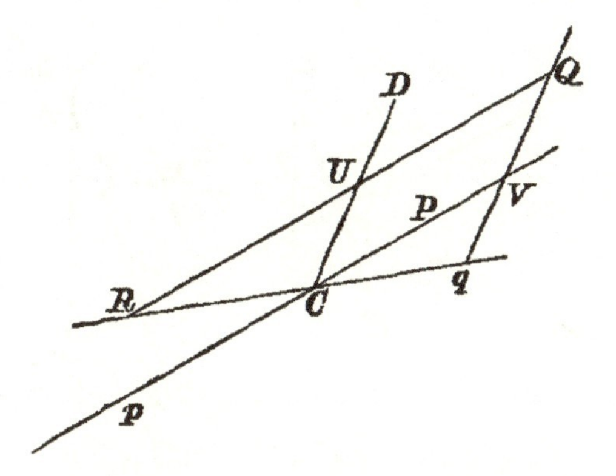

Hence, when two diameters are conjugate, each bisects the chords parallel to the other.

DEF. *Chords drawn from the extremities of any diameter to a point on the hyperbola are called supplemental chords.*

Thus, qQ, QR are supplemental chords, and they are parallel to CD and CP; supplemental chords are therefore parallel to conjugate diameters.

DEF. *A line QV, drawn from any point Q of an hyperbola, parallel to a diameter DCd, and terminated by the conjugate diameter PCp, is called an ordinate of the diameter PCp, and if QV produced meet the curve in Q', QVQ' is the double ordinate.*

This definition includes the two cases in which QQ' may be drawn so as to meet the same, or opposite branches of the hyperbola.

120. PROP. XIX. *Any diameter is a mean proportional between the transverse axis and the focal chord parallel to the diameter.*

This can be proved as in Art. 81.

Properties of Asymptotes.

121. PROP. XX. *If from any point Q in an asymptote $QPpq$ be drawn meeting the curve in P, p and the other asymptote in q, and if CD be the semi-diameter parallel to Qq,*

$$QP \,.\, Pq = CD^2 \text{ and } QP = pq.$$

Through P and D draw RPr, DTt perpendicular to the transverse axis, and meeting the asymptotes.

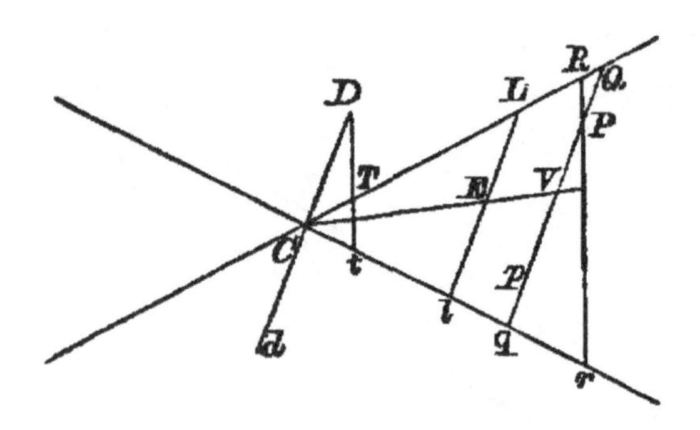

Then $$QP : RP :: CD : DT,$$

and $$Pq : Pr :: CD : Dt;$$

$$\therefore QP \,.\, Pq : RP \,.\, Pr :: CD^2 : DT \,.\, Dt.$$

But $$RP \,.\, Pr = BC^2 = DT \,.\, Dt \text{ (Arts. 106 and 107)},$$

$$\therefore QP \,.\, Pq = CD^2.$$

Similarly $$qp \,.\, pQ = CD^2;$$

$$\therefore QP \,.\, Pq = qp \,.\, pQ;$$

or, if V be the middle point of Qq,

$$QV^2 - PV^2 = QV^2 - pV^2.$$

Hence $$PV = pV, \text{ and } \therefore PQ = pq.$$

We have taken the case in which Qq meets one branch of the hyperbola. It may however be shewn in the same manner that the same relations hold good for the case in which Qq meets opposite branches.

COR. *If a straight line $PP'p'p$ meet the hyperbola in P, p, and the conjugate hyperbola in P', p', $PP' = pp'$.*

For, if the line meet the asymptotes in Q, q,

$$QP' = p'q, \text{ and } PQ = qp;$$

$$\therefore PP' = pp'.$$

122. PROP. XXI. *The portion of a tangent which is terminated by the asymptotes is bisected at the point of contact, and is equal to the parallel diameter.*

LEl being the tangent (Fig. Art. 121), and DCd the parallel diameter, draw any parallel straight line $QPpq$ meeting the curve and the asymptotes.

Then $QP = pq$; and, if the line move parallel to itself until it coincides with Ll, the points P and p coincide with E, and $\therefore LE = El$.

Also $$QP \,.\, Pq = CD^2, \text{ always;}$$
$$\therefore LE \,.\, El = CD^2, \text{ or } LE = CD.$$

Properties of Conjugate Diameters.

123. Prop. XXII. *Conjugate diameters of an hyperbola are also conjugate diameters of the conjugate hyperbola, and the asymptotes are diagonals of the parallelogram formed by the tangents at their extremities.*

PCp and DCd being conjugate, let QVq, a double ordinate of CD, meet the conjugate hyperbola in Q' and q'.

Then $$QV = Vq, \text{ and } QQ' = qq' \text{ (Cor. Art. 121)},$$
$$\therefore Q'V = Vq'.$$

That is, CD bisects the chords of the conjugate hyperbola parallel to CP.

Hence CD and CP are conjugate in both hyperbolas, and therefore the tangent at D is parallel to CP.

Let the tangent at P meet the asymptote in L; then
$$PL = CD \text{ (Art. 122)}.$$

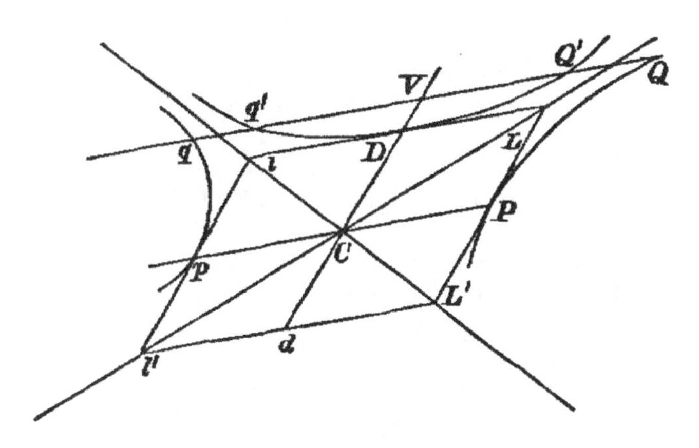

Hence LD is parallel and equal to CP;
but the tangent at D is parallel to CP;
$$\therefore LD \text{ is the tangent at } D.$$

Completing the figure, the tangents at p and d are parallel to those at P and D, and therefore the asymptotes are the diagonals of the parallelogram $Lll'L'$.

COR. Hence, joining PD, it follows that PD is parallel to the asymptote lCL', since $LP = PL'$, and $LD = Dl$,

124. PROP. XXIII. *If QV be an ordinate of a diameter PCp, and DCd the conjugate diameter,*

$$QV^2 : PV \,.\, Vp :: CD^2 : CP^2.$$

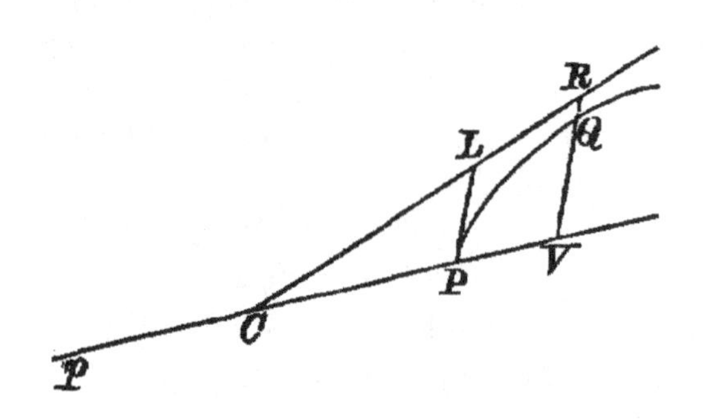

Let QV and the tangent at P meet the asymptote in R and L.

Then LP being equal to CD,

$$RV^2 : CD^2 :: CV^2 : CP^2;$$

$$\therefore RV^2 - CD^2 : CD^2 :: CV^2 - CP^2 : CP^2.$$

But $$RV^2 - QV^2 = CD^2.$$

Hence $$QV^2 : CD^2 :: CV^2 - CP^2 : CP^2,$$

$$\text{or } QV^2 : PV \,.\, Vp :: CD^2 : CP^2.$$

125. PROP. XXIV. *If QV be an ordinate of a diameter PCp, and if the tangent at Q meet the conjugate diameter, DCd, in t,*

$$Ct \,.\, QV = CD^2.$$

For, (Fig. Art. 118)

$$Ct : QV :: CT : VT,$$

$$\text{and } \therefore Ct \,.\, QV : QV^2 :: CV \,.\, CT : CV \,.\, VT.$$

But $$CV \,.\, CT = CP^2,$$

$$\text{and } CV \,.\, VT = CV^2 - CV \,.\, CT = CV^2 - CP^2;$$

$$\therefore Ct \,.\, QV : QV^2 :: CP^2 : CV^2 - CP^2,$$

$$:: CD^2 : QV^2.$$

Hence $$Ct \,.\, QV = CD^2.$$

126. Prop. XXV. *If ACa, BCb be conjugate diameters, and PCp, DCd another pair of conjugate diameters, and if PN, DM be ordinates of ACa,*

$$CM : PN :: AC : BC,$$

$$\text{and } DM : CN :: BC : AC.$$

Let the tangents at P and D meet ACa in T and t; then $CN \,.\, CT = AC^2 = CM \,.\, Ct$ (Art. 117),

$$\therefore CM : CN :: CT : Ct,$$

$$:: PT : CD,$$

$$:: PN : DM,$$

$$:: CN : Mt;$$

$$\therefore CN^2 = CM \,.\, Mt = CM^2 + CM \,.\, Ct = CM^2 + AC^2,$$

so that $$CM^2 = CN^2 - AC^2.$$

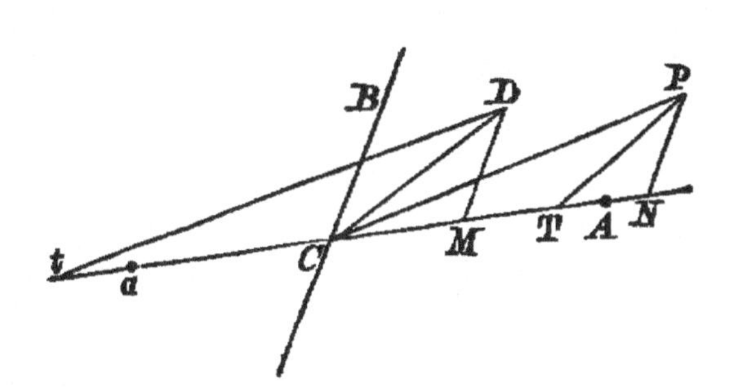

But $$PN^2 : CN^2 - AC^2 :: BC^2 : AC^2;$$

$$\therefore CM : PN :: AC : BC;$$

and, similarly, $$DM : CN :: BC : AC.$$

Cor. We have shewn in the course of the proof, that

$$CN^2 - CM^2 = AC^2.$$

Similarly, if Pn, Dm be ordinates of BC,

$$Cm^2 - Cn^2 = BC^2;$$

that is, $$DM^2 - PN^2 = BC^2;$$

and it must be noticed that these relations are shewn for any pair of conjugate diameters ACa, BCb, including of course the axes.

127. PROP. XXVI. *If CP, CD be conjugate semi-diameters, and AC, BC the semi-axes,*

$$CP^2 - CD^2 = AC^2 - BC^2.$$

For, drawing the ordinates PN, DM, and remembering that in this case the angles at N and M are right angles, we have, from the figure of the previous article,

$$CP^2 = CN^2 + PN^2,$$
$$CD^2 = CM^2 + DM^2.$$

But $CN^2 - CM^2 = AC^2$ and $DM^2 - PN^2 = BC^2$;

$$\therefore CP^2 - CD^2 = AC^2 - BC^2.$$

128. PROP. XXVII. *If the normal at P meet the axes in G and g,*

$$PG : CD :: BC : AC,$$

and

$$Pg : CD :: AC : BC.$$

For the proofs of these relations, see Art. 86.

Observe also that

$$PG \,.\, Pg = CD^2,$$

and that

$$Gg : CD :: SC^2 : AC \,.\, BC.$$

129. PROP. XXVIII. *The area of the parallelogram formed by the tangents at the ends of conjugate diameters is equal to the rectangle contained by the axes.*

Let CP, CD be the semi-diameters, and PN, DM the ordinates of the transverse axis.

Let the normal at P meet CD in F, and the axis in G. Then PNG, CDM are similar triangles, and, exactly as in Art. 87, it can be shewn that

$$PF \,.\, CD = AC \,.\, BC.$$

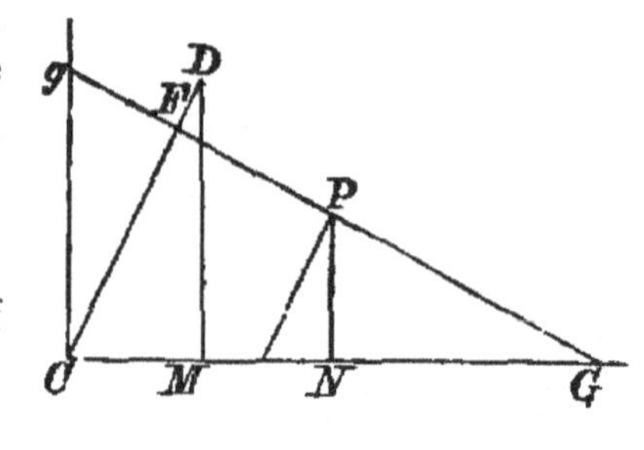

Hence it follows that, in the figure of Art. 123, the triangle LCL' is of constant area.

For the triangle is equal to the parallelogram $CPLD$.

130. Prop. XXIX. *If SP, $S'P$ be the focal distances of a point P, and CD be conjugate to CP,*

$$SP \,.\, S'P = CD^2.$$

Attending to the figure of Art. 111, the proof is the same as that of Art. 88.

131. Prop. XXX. *If the tangent at P meet a pair of conjugate diameters in T and t, and CD be conjugate to CP,*

$$PT \,.\, Pt = CD^2.$$

This can be proved as in Art. 89.

It can also be shewn that if the tangent at P meet two parallel tangents in T' and t',

$$PT' \,.\, Pt' = CD^2.$$

132. Prop. XXXI. *If the tangent at P meet the asymptotes in L and L',*

$$CL \,.\, CL' = SC^2.$$

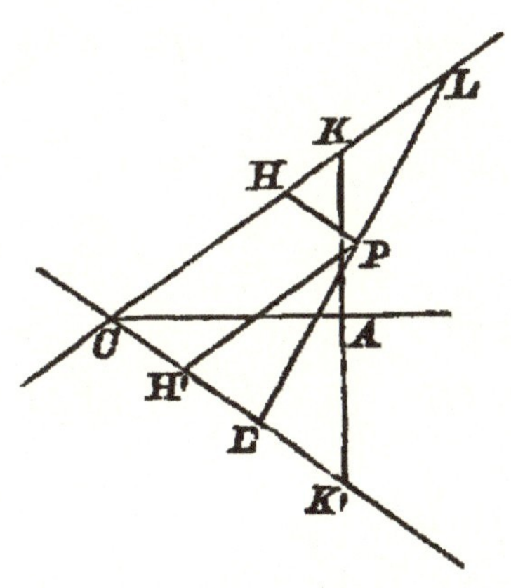

Let the tangent at A meet the asymptotes in K and K'; then (Art. 129) the triangles LCL', KCK' are of equal area, and therefore

$$CL : CK' :: CK : CL' \text{ (Euclid, Book VI.)},$$

or

$$CL \,.\, CL' = CK^2 = AC^2 + BC^2 = SC^2.$$

Cor. If PH, PH' be drawn parallel to, and terminated by the asymptotes,

$$4 \,.\, PH \,.\, PH' = CS^2,$$

for $CL = 2PH'$, and $CL' = 2PH$.

133. Prop. XXXII. *Pairs of tangents at right angles to each other intersect on a fixed circle.*

PT, QT being two tangents at right angles, let SY, perpendicular to PT, meet $S'P$ in K.

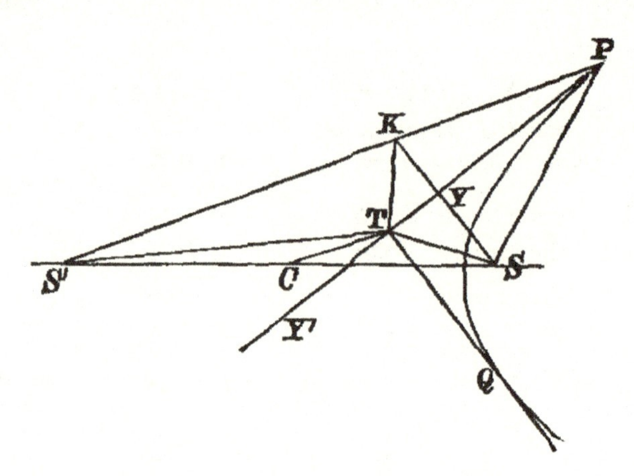

Then (Art. 113) the angle $S'TY' = QTS$,

and obviously, $KTP = PTS$;

therefore $S'TY'$ is complementary to KTP, and $S'TK$ is a right angle.

Hence

$$4AC^2 = S'K^2 = S'T^2 + TK^2$$
$$= S'T^2 + ST^2$$
$$= 2 . CT^2 + 2 . CS^2 \text{ by Euclid II. 12 and 13;}$$
$$\therefore CT^2 = AC^2 - BC^2,$$

and the locus of T is a circle.

If AC be less than BC, this relation is impossible.

In this case, however, the angle between the asymptotes is greater than a right angle, and the angle PTQ between a pair of tangents, being always greater than the angle between the asymptotes, is greater than a right angle. The problem is therefore *à priori* impossible for the hyperbola, but becomes possible for the conjugate hyperbola.

As in the case of the ellipse, the locus of T is called the director circle.

134. Prop. XXXIII. *The rectangles contained by the segments of any two chords which intersect each other are in the ratio of the squares on the parallel diameters.*

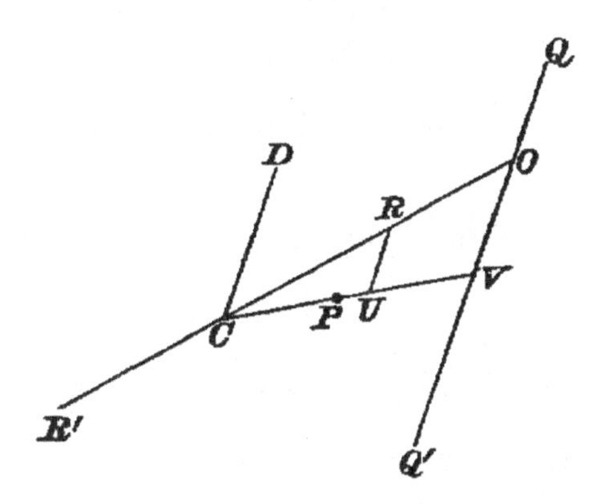

Through any point O in a chord QOQ' draw the diameter ORR'; and let CD be parallel to QQ', and CP conjugate to CD, bisecting QQ' in V.

Draw RU an ordinate of CP.

Then $$RU^2 : CU^2 - CP^2 :: CD^2 : CP^2;$$
$$\therefore CD^2 + RU^2 : CU^2 :: CD^2 : CP^2,$$
$$:: CD^2 + QV^2 : CV^2.$$

But $$RU^2 : CU^2 :: OV^2 : CV^2;$$

$$\therefore CD^2 : CU^2 :: CD^2 + QV^2 - OV^2 : CV^2,$$

or $$CD^2 : CD^2 + QV^2 - OV^2 :: CU^2 : CV^2,$$
$$:: CR^2 : CO^2;$$

$$\therefore CD^2 : QV^2 - OV^2 :: CR^2 : CO^2 - CR^2,$$

or $$CD^2 : QO \,.\, OQ' :: CR^2 : OR \,.\, OR'.$$

Similarly, if qOq' be any other chord, and Cd the parallel semi-diameter,
$$Cd^2 : qO \,.\, Oq' :: CR^2 : OR \,.\, OR';$$
$$\therefore QO \,.\, OQ' : qO \,.\, Oq' :: CD^2 : Cd^2;$$

that is, the ratio of the rectangles depends only on the directions of the chords.

PROP. XXXIV. *If a circle intersect an hyperbola in four points, the several pairs of the chords of intersection are equally inclined to the axes.*

For the proof, see Art. 93.

EXAMPLES.

1. If a circle be drawn so as to touch two fixed circles externally, the locus of its centre is an hyperbola.

2. If the tangent at B to the conjugate meet the latus rectum in D, the triangles SCD, SXD are similar.

3. The straight line drawn from the focus to the directrix, parallel to an asymptote, is equal to the semi-latus rectum, and is bisected by the curve.

4. Given the asymptotes and a focus, find the directrix.

5. Given the centre, one asymptote, and a directrix, find the focus.

6. Parabolas are described passing through two fixed points, and having their axes parallel to a fixed line; the locus of their foci is an hyperbola.

7. The base of a triangle being given, and also the point of contact with the base of the inscribed circle, the locus of the vertex is an hyperbola.

8. If the normal at P meet the conjugate axis in g, and gN be the perpendicular on SP, then $PN = AC$.

9. Draw a tangent to an hyperbola, or its conjugate, parallel to a given line.

10. If AA' be the axis of an ellipse, and PNP' a double ordinate, the locus of the intersection of $A'P$ and $P'A$ is an hyperbola.

11. The tangent at P bisects any straight line perpendicular to AA', and terminated by AP, and $A'P$.

12. If PCp be a diameter, and if Sp meet the tangent at P in T,

$$SP = ST.$$

13. Given an asymptote, the focus, and a point; construct the hyperbola.

14. A circle can be drawn through the foci and the intersections of any tangent with the tangents at the vertices.

15. Given an asymptote, the directrix, and a point; construct the hyperbola.

16. If through any point of an hyperbola straight lines are drawn parallel to the asymptotes and meeting any semi-diameter CQ in P and R,

$$CP \,.\, CR = CQ^2.$$

17. PN is an ordinate and NQ parallel to AB meets the conjugate axis in Q; prove that $QB \,.\, QB' = PN^2$.

18. NP is an ordinate and Q a point in the curve; AQ, $A'Q$ meet NP in D and E; prove that $ND \,.\, NE = NP^2$.

19. If a tangent cut the major axis in the point T, and perpendiculars SY, HZ be let fall on it from the foci, then

$$AT \,.\, A'T = YT \,.\, ZT.$$

20. In the tangent at P a point Q is taken such that PQ is proportional to CD; shew that the locus of Q is an hyperbola.

21. Tangents are drawn to an hyperbola, and the portion of each tangent intercepted by the asymptotes is divided in a constant ratio; prove that the locus of the point of section is an hyperbola.

22. If the tangent and normal at P meet the conjugate axis in t and K respectively, prove that a circle can be drawn through the foci and the three points P, t, K.

Shew also that

$$GK : SK :: SA : AX,$$

and

$$St : tK :: BC : CD,$$

CD being conjugate to CP.

23. Shew that the points of trisection of a series of conterminous circular arcs lie on branches of two hyperbolas; and determine the distance between their centres.

24. If the tangent at any point P cut an asymptote in T, and if SP cut the same asymptote in Q, then $SQ = QT$.

25. A series of hyperbolas having the same asymptotes is cut by a straight line parallel to one of the asymptotes, and through the points of intersection lines are drawn parallel to the other, and equal to either semi-axis of the corresponding hyperbola: prove that the locus of their extremities is a parabola.

26. Prove that the rectangle $PY \,.\, PY'$ in an ellipse is equal to the square on the conjugate axis of the confocal hyperbola passing through P.

27. If the tangent at P meet one asymptote in T, and a line TQ be drawn parallel to the other asymptote to meet the curve in Q; prove that if PQ be joined and produced both ways to meet the asymptotes in R and R', RR' will be trisected at the points P and Q.

28. The tangent at a point P of an ellipse meets the hyperbola having the same axes as the ellipse in C and D. If Q be the middle point of CD, prove that OQ and OP are equally inclined to the axes, O being the centre of the ellipse.

29. Given one asymptote, the direction of the other, and the position of one focus, determine the position of the vertices.

30. Two points are taken on the same branch of the curve, and on the same side of the axis; prove that a circle can be drawn touching the four focal distances.

31. Supposing the two asymptotes and one point of the curve to be given in position, shew how to construct the curve; and find the position of the foci.

32. Given a pair of conjugate diameters, construct the axes.

33. If PH, PK be drawn parallel to the asymptotes from a point P on the curve, and if a line through the centre meet them in R, T, and the parallelogram $PRQT$ be completed, Q is a point on the curve.

34. The ordinate NP at any point of an ellipse is produced to a point Q, such that NQ is equal to the sub-tangent at P; prove that the locus of Q is an hyperbola.

35. If a given point be the focus of any hyperbola, passing through a given point and touching a given straight line, prove that the locus of the other focus is an arc of a fixed hyperbola.

36. An ellipse and hyperbola are described, so that the foci of each are at the extremities of the transverse axis of the other; prove that the tangents at their points of intersection meet the conjugate axis in points equidistant from the centre.

37. A circle is described about the focus as centre, with a radius equal to one-fourth of the latus rectum; prove that the focal distances of the points at which it intersects the hyperbola are parallel to the asymptotes.

38. The tangent at any point forms a triangle with the asymptotes: determine the locus of the point of intersection of the straight lines drawn from the angles of this triangle to bisect the opposite sides.

39. If SY, $S'Y'$ be the perpendiculars on the tangent at P, a circle can be drawn through the points Y, Y', N, C.

40. The straight lines joining each focus to the foot of the perpendicular from the other focus on the tangent meet on the normal and bisect it.

41. If the tangent and normal at P meet the axis in T and G, $NG.CT = BC^2$.

42. If the tangent at P meet the axes in T and t, the angles PSt, STP are supplementary.

43. If the tangent at P meet any conjugate diameters in T and t, the triangles SPT, $S'Pt$ are similar.

44. If the diameter conjugate to CP meet SP and $S'P$ in E and E', prove that the circles about the triangles SCE, $S'CE'$ are equal.

45. The locus of the centre of the circle inscribed in the triangle SPS' is a straight line.

46. If PN be an ordinate, and NQ parallel to AP meet CP in Q, AQ is parallel to the tangent at P.

47. If an asymptote meet the directrix in D, and the tangent at the vertex in E, AD is parallel to SE.

48. The radius of the circle touching the curve and its asymptotes is equal to the portion of the latus rectum produced, between its extremity and the asymptote.

49. If G be the foot of the normal, and if the tangent meet the asymptotes in L and M, $GL = GM$.

50. With two conjugate diameters of an ellipse as asymptotes, a pair of conjugate hyperbolas is constructed: prove that if one hyperbola touch the ellipse, the other will do so likewise; prove also that the diameters drawn through the points of contact are conjugate to each other.

51. If two tangents be drawn the lines joining their intersections with the asymptotes will be parallel.

52. The locus of the centre of the circle touching SP, $S'P$ produced, and the major axis, is an hyperbola.

53. If from a point P in an hyperbola, PK be drawn parallel to an asymptote to meet the directrix in K, then $PK = SP$.

54. If PD be drawn parallel to an asymptote, to meet the conjugate hyperbola in D, CP and CD are conjugate diameters.

55. If QR be a chord parallel to the tangent at P, and if QL, PN, RM be drawn parallel to one asymptote to meet the other,

$$CL \,.\, CM = CN^2.$$

56. If a circle touch the transverse axis at a focus, and pass through one end of the conjugate, the chord intercepted by the conjugate is a third proportional to the conjugate and transverse semi-axes.

57. A line through one of the vertices, terminated by two lines drawn through the other vertex parallel to the asymptotes, is bisected at the other point where it cuts the curve.

58. If PSQ be a focal chord, and if the tangents at P and Q meet in T, the difference between PTQ and half $PS'Q$ is a right angle.

59. If a straight line passing through a fixed point C meet two fixed lines OA, OB in A and B, and if P be taken in AB such that $CP^2 = CA \,.\, CB$, the locus of P is an hyperbola, having its asymptotes parallel to OA, OB.

60. If from the points P and Q in an hyperbola there be drawn PL, QM parallel to each other to meet one asymptote, and PR, QN also parallel to each other to meet the other asymptote, $PL \,.\, PR = QM \,.\, QN$.

61. Prove that the locus of the point of intersection of two tangents to a parabola which cut at a constant angle is an hyperbola, and that the angle between its asymptotes is double the external angle between the tangents.

62. An ordinate VQ of any diameter CP is produced to meet the asymptote in R, and the conjugate hyperbola in Q'; prove that

$$QV^2 + Q'V^2 = 2RV^2.$$

Prove also that the tangents at Q and Q' meet the diameter CP in points equidistant from C.

63. A chord QPL meets an asymptote in L, and a tangent from L is drawn touching at R; if PM, RE, QN, be drawn parallel to the asymptote to meet the other,

$$PM + QN = 2 \,.\, RE.$$

64. Tangents are drawn from any point in a circle through the foci; prove that the lines bisecting the angle between the tangents, or between one tangent and the other produced, all pass through a fixed point.

65. If a circle through the foci meet two confocal hyperbolas in P and Q, the angle between the tangents at P and Q is equal to PSQ.

66. If SY, $S'Y'$ be perpendiculars on the tangent at P, and if PN be the ordinate, the angles PNY, PNY' are supplementary.

67. Find the position of P when the area of the triangle YCY' is the greatest possible, and shew that, in that case,

$$PN \,.\, SC = BC^2.$$

68. If the tangent at P meet the conjugate axis in t, the areas of the triangles SPS', StS' are in the ratio of $CD^2 : St^2$.

69. If SY, SZ be perpendiculars on two tangents which meet in T, YZ is perpendicular to $S'T$.

70. A circle passing through a focus, and having its centre on the transverse axis, touches the curve; shew that the focal distance of the point of contact is equal to the latus rectum.

71. If CQ be conjugate to the normal at P, then is CP conjugate to the normal at Q.

72. From a point in the auxiliary circle lines are drawn touching the curve in P and P'; prove that SP, $S'P'$ are parallel.

73. If any hyperbola is drawn confocal with a given ellipse, and if PN is the ordinate of a point of intersection of the hyperbola with the ellipse, and NT the tangent from N to the auxiliary circle of the hyperbola, prove that the angle PNT is always the same.

74. Find the locus of the points of contact of tangents to a series of confocal hyperbolas from a fixed point in the axis.

75. Tangents to an hyperbola are drawn from any point in one of the branches of the conjugate, shew that the chord of contact will touch the other branch of the conjugate.

76. An ordinate NP meets the conjugate hyperbola in Q; prove that the normals at P and Q meet on the transverse axis.

77. A parabola and an hyperbola have a common focus S and their axes in the same direction. If a line SPQ cut the curves in P and Q, the angle between the tangents at P and Q is equal to half the angle between the axis and the other focal distance of the hyperbola.

78. If an hyperbola be described touching the four sides of a quadrilateral which is inscribed in a circle, and one focus lie on the circle, the other focus will also lie on the circle.

79. A conic section is drawn touching the asymptotes of an hyperbola. Prove that two of the chords of intersection of the curves are parallel to the chord of contact of the conic with the asymptotes.

80. A parabola P and an hyperbola H have a common focus, and the asymptotes of H are tangents to P; prove that the tangent at the vertex of P is a directrix of H, and that the tangent to P at the point of intersection passes through the further vertex of H.

81. From a given point in an hyperbola draw a straight line such that the segment intercepted between the other intersection with the hyperbola and a given asymptote shall be equal to a given line.

When does the problem become impossible?

82. If an ellipse and a confocal hyperbola intersect in P, an asymptote passes through the point on the auxiliary circle of the ellipse corresponding to P.

83. P is a point on an hyperbola whose foci are S and H; another hyperbola is described whose foci are S and P, and whose transverse axis is equal to $SP-2PH$: shew that the hyperbolas will meet only at one point, and that they will have the same tangent at that point.

84. A point D is taken on the axis of an hyperbola, of which the eccentricity is 2, such that its distance from the focus S is equal to the distance of S from the further vertex A'; P being any point on the curve, $A'P$ meets the latus rectum in K. Prove that DK and SP intersect on a certain fixed circle.

85. Shew that the locus of the point of intersection of tangents to a parabola, making with each other a constant angle equal to half a right angle, is an hyperbola.

86. The tangent and normal at any point intersect the asymptotes and axes respectively in four points which lie on a circle passing through the centre of the curve.

The radius of this circle varies inversely as the perpendicular from the centre on the tangent.

87. The difference between the sum of the squares of the distances of any point from the ends of any diameter and the sum of the squares of its distances from the ends of the conjugate is constant.

88. If a tangent meet the asymptotes in L and M, the angle subtended by LM at the farther focus is half the angle between the asymptotes.

89. If PN be the ordinate of P, and PT the tangent, prove that $SP : ST :: AN : AT$.

90. If an ellipse and an hyperbola are confocal, the asymptotes pass through the points on the auxiliary circle of the ellipse which correspond to the points of intersection of the two curves.

91. Two adjacent sides of a quadrilateral are given in magnitude and position; if the quadrilateral be such that a circle can be inscribed in it, the locus of the point of intersection of the other two sides is an hyperbola.

92. The tangent at P meets the conjugate axis in t, and tQ is perpendicular to SP; prove that SQ is of constant length.

93. An hyperbola, having a given transverse axis, has one focus fixed, and always touches a given straight line; the locus of the other focus is a circle.

94. A chord $PRVQ$ meets the directrices in R and V; shew that PR and VQ subtend, each at the focus nearer to it, angles of which the sum is equal to the angle between the tangents at P and Q.

95. A circle is drawn touching the transverse axis of an hyperbola at its centre, and also touching the curve; prove that the diameter conjugate to the diameter through either point of contact is equal to the distance between the foci.

96. A parabola is described touching the conjugate axes of an hyperbola at their extremities; prove that one asymptote is parallel to the axis of the parabola, and that the other asymptote is parallel to the chords of the parabola bisected by the first.

If a straight line parallel to the second asymptote meet the hyperbola and its conjugate in P, P', and the parabola in Q, Q', it may be shewn that $PQ = P'Q'$.

97. If two points E and E' be taken in the normal PG such that $PE = PE' = CD$, the loci of E and E' are hyperbolas having their axes equal to the sum and difference of the axes of the given hyperbola.

98. If two tangents are drawn to the same branch of an hyperbola, the external angle between them is half the difference between the angles which the chord of contact subtends at the foci.

If the tangents are drawn to opposite branches, the angle between them is half the sum, or half the difference, of these angles according as the points of contact are on the same or on opposite sides of the transverse axis.

99. Parabolas are drawn passing through two fixed points A and B, and having their axes in a given direction; find the locus of the foci, and, if a tangent be drawn at right angles to AB, prove that the locus of its point of contact P is an hyperbola.

100. Tangents are drawn from a point T to an hyperbola whose centre is C, and CT produced meets the hyperbola in P and the chord of contact of the tangents in V. If CD be the diameter conjugate to CP, and DT, DV meet the tangent at P in K and U, prove that the triangles PUV, TPK are equal in area.

101. One asymptote and three points P, Q, R of an hyperbola are given, construct the other asymptote.

102. If an ellipse be described having its centre on a given hyperbola, its foci on the asymptotes, and passing through the centre of the hyperbola, prove that the minor axis of the ellipse is equal to the major axis of the hyperbola, and the ellipse touches the minor axis of the hyperbola.

103. The angular point A of a triangle ABC is fixed, and the angle A is given, while the points B and C move on a fixed straight line; prove that the locus of the centre of the circle circumscribing the triangle is an hyperbola, and that the envelope of the circle is another circle.

104. Given an asymptote CQ and two points on an hyperbola, P, p on the curve, shew that the envelope of the axes is a parabola.

105. Find the locus of the middle points of a system of chords of an hyperbola, passing through a fixed point on one of the asymptotes.

106. If a conic be described having for its axes the tangent and normal at any point of a given ellipse, and touching at its centre the axis-major of the given ellipse, and if another conic be described in the same manner, but touching the minor axis at the centre, prove that the foci of these conics lie in two circles concentric with the given ellipse, and having their diameters equal to the sum and difference of its axes.

107. An ellipse and an hyperbola are confocal; if a tangent to one intersect at right angles a tangent to the other, the locus of the point of intersection is a circle.

Shew also that the difference of the squares on the distances from the centre of parallel tangents is constant.

108. If a circle passing through any point P of the curve, and having its centre on the normal at P, meets the curve again in Q and R, the tangents at Q and R intersect on a fixed straight line.

109. If the tangent at P meet an asymptote in T, the angle between that asymptote and $S'P$ is double the angle STP.

110. Four tangents to an hyperbola form a rectangle. If one side AB of the rectangle intersect a directrix in F, and S be the corresponding focus, the triangles FSA, FBS are similar.

111. An ellipse and hyperbola have the same transverse axis, and their eccentricities are the reciprocals of one another; prove that the tangents to each through the focus of the other intersect at right angles in two points and also meet the conjugate axis on the auxiliary circle.

112. ACA' and BCB' are the transverse and conjugate axes of an ellipse, of which S and S' are the foci. P is one of the points of intersection of this ellipse and a confocal hyperbola, and aCa' is the transverse axis of the hyperbola.

Prove that $SP = Aa$, $S'P = A'a$, and $aB = CP$.

113. Prove that if A, B and S are three given points, two parabolas can be drawn through A and B with S as focus, and that the axes of these parabolas are parallel to the asymptotes of the hyperbola which can be drawn through S with its foci at A and B.

CHAPTER V.

THE RECTANGULAR HYPERBOLA.

If the axes of an hyperbola be equal, the angle between the asymptotes is a right angle, and the curve is called equilateral or rectangular.

135. PROP. I. *In a rectangular hyperbola*

$$CS^2 = 2AC^2, \text{ and } SA^2 = 2AX^2.$$

For
$$CS^2 = AC^2 + BC^2 = 2AC^2,$$
and
$$SA : AX :: SC : AC;$$
$$\therefore SA^2 = 2AX^2.$$

Observe that, in the figure of Art. 102, SDC is an isosceles triangle, since

$$SD = BC, \text{ and } CD = AC,$$

and therefore
$$SD = DC.$$

136. PROP. II. *The asymptotes of a rectangular hyperbola bisect the angles between any pair of conjugate diameters.*

For, in a rectangular or equilateral hyperbola,

$$CA = CB,$$

and therefore, since
$$CP^2 - CD^2 = CA^2 - CB^2,$$
$$CP = CD,$$

CP, CD being any conjugate semi-diameters.

Also, in the figure of Art. 123, the parallelogram $CPLD$ is a rhombus, and therefore CL bisects the angle PCD.

COR. Supplemental chords are equally inclined to the asymptotes, for they are parallel to conjugate diameters.

137. Prop. III. *If CY be the perpendicular from the centre on the tangent at P, the angle PCY is bisected by the transverse axis, and half the transverse axis is a mean proportional between CY and CP.*

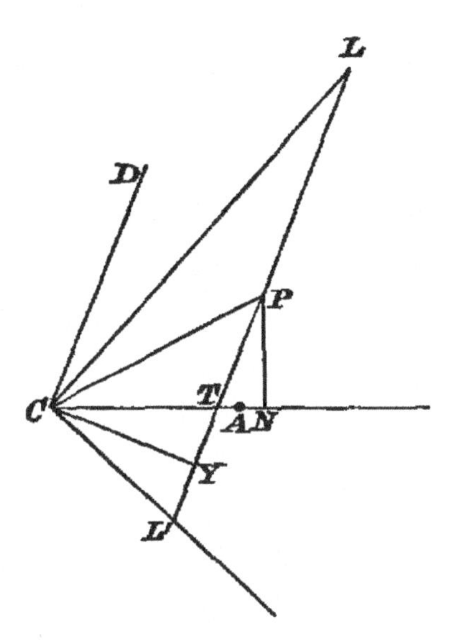

For the angle $PCL = DCL$

$$= YCL';$$

$$\therefore PCA = ACY.$$

Hence it follows that the triangles PCN, TCY are similar, and that

$$CY : CT :: CN : CP;$$

$$\therefore CY \,.\, CP = CT \,.\, CN = AC^2.$$

Hence also, if we join PA and AY, we observe that the triangles PAC, AYC are similar.

138. Prop. IV. *Diameters at right angles to each other are equal.*

Let CP, CP' be semi-diameters at right angles to each other, and CD conjugate to CP.

Then, if CL, CL' be the asymptotes,

$$\text{the angle } P'CL' = PCL = DCL;$$

$$\therefore CP' = CD = CP.$$

Hence it follows, by help of the theorem of Art. 120, that focal chords at right angles to each other are equal, and that focal chords parallel to conjugate diameters are equal.

139. Prop. V. *If the normal at P meet the axes in G and g,*

$$CN = NG \text{ and } PG = Pg = CD,$$

CD being conjugate to CP.

For (Art. 115) $NG : CN :: BC^2 : AC^2$;

$$\therefore NG = CN.$$

Also $\quad PF \,.\, PG = BC^2$ and $PF \,.\, Pg = AC^2$;

$$\therefore PG = Pg.$$

Further (Art. 128) $PG : CD :: BC : AC$;

$$\therefore PG = CD = CP.$$

140. PROP. VI. *If QV be an ordinate of a diameter PCp,*

$$QV^2 = PV \,.\, Vp.$$

For
$$QV^2 : PV \,.\, Vp :: CD^2 : CP^2,$$

and
$$CD = CP;$$
$$QV^2 = PV \,.\, Vp = CV^2 - CP^2.$$

141. PROP. VII. *The angle between a chord PQ, and the tangent at P, is equal to the angle subtended by PQ at the other extremity of the diameter through P.*

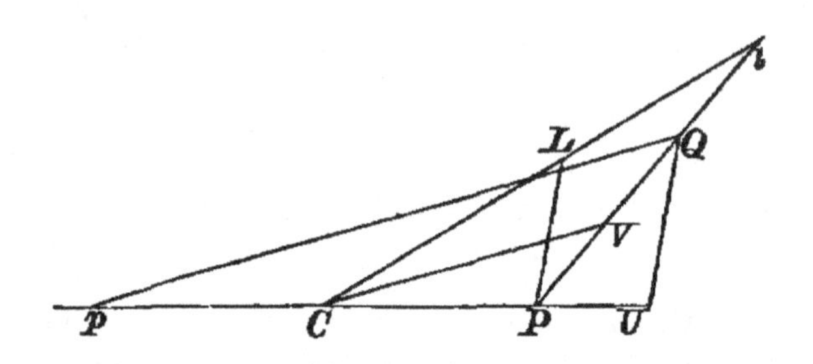

Let PQ and the tangent at P meet the asymptote in l and L. Then, if CV be conjugate to PQ,

$$\text{the angle } LPQ = PLC - VlC = LCP - VCl$$
$$= VCP = QpP.$$

Or thus, let QU parallel to the tangent at P, meet CP produced in U. Then

$$QU^2 = PU \,.\, Up,$$

or,

$$QU : PU :: Up : UQ.$$

Therefore the triangles PQU, QUp are similar, and the angle $QpU = PQU = LPQ$.

If P and Q are on opposite branches of the curve, the same proof shews that

$$\text{the angle } QpU = UQP = LPQ;$$

$$\therefore QPL' = QpP.$$

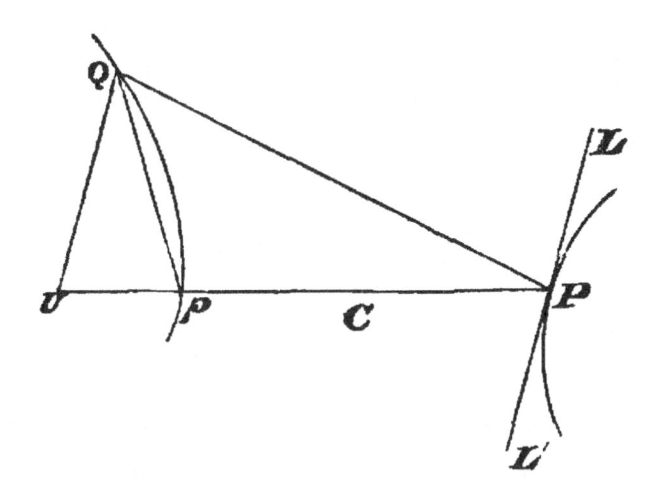

If QP is the normal at P, it follows that QP subtends a right angle at the other end of the diameter through P.

142. PROP. VIII. *Any chord subtends, at the ends of any diameter, angles which are equal or supplementary.*

This theorem divides itself into four cases, which are shewn in the appended figures.

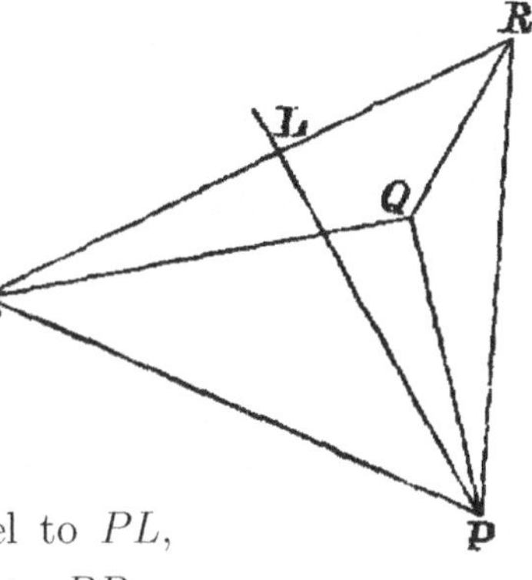

Let QR be the chord, and Pp the diameter. Then, if LP be the tangent at P, fig. (1),

$$\text{the angle } LPQ = QpP,$$
$$\text{and } LPR = RpP;$$
$$\therefore QPR = QpR.$$

In fig. (2), if pl be the tangent at p, parallel to PL,

$$QpR = Qpl + lpR = Qpl + pPR,$$

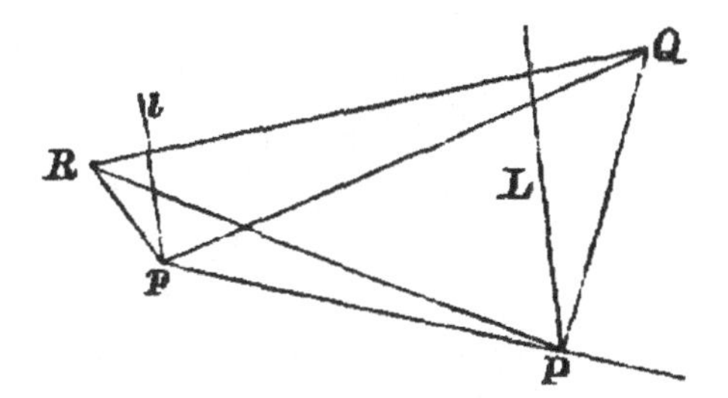

and $$QPR = QPL + LPR = QpP + LPR;$$
$$\therefore QpR + QPR = lpP + LPp,$$
that is, QpR and QPR are together equal to two right angles.

In fig. (3)
$$QPR = QPL + LPp + pPR$$
$$= QpP + Ppl + lpR$$
$$= QpR.$$

In fig. (4) $QPL = QpP$, and $RPL' = RpP$;
$$\therefore QpR = QPL + RPL';$$
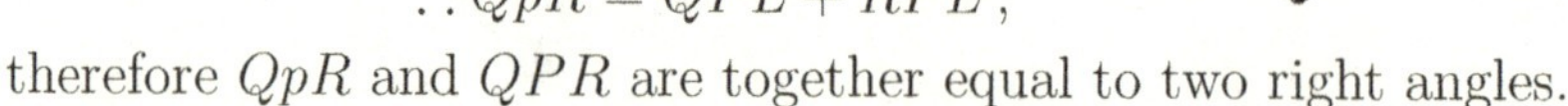
therefore QpR and QPR are together equal to two right angles.

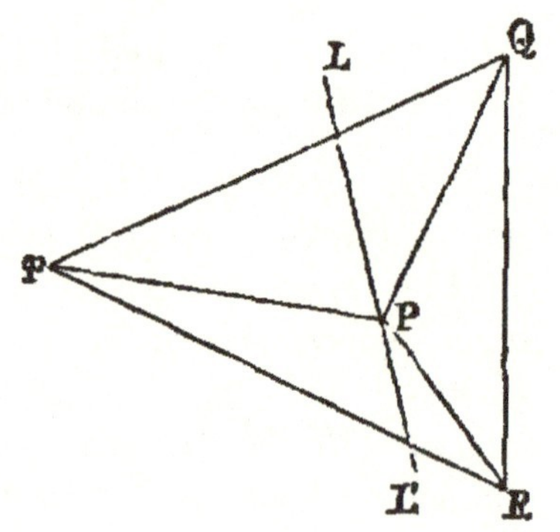

Hence it will be seen that when QR, or QR produced, meet the diameter Pp between P and p, the angles subtended at P and p are equal; in other cases they are supplementary.

In the cases of the second and third figures, if one of the angles QPR is a right angle, the other angle QpR is also a right angle. The four points Q, P, p, R are then concyclic, and QR is a diameter of the circle.

143. PROP. IX. *If a rectangular hyperbola circumscribe a triangle, it passes through the orthocentre.*

NOTE. *The orthocentre is the point of intersection of the perpendiculars from the angular points on the opposite sides.*

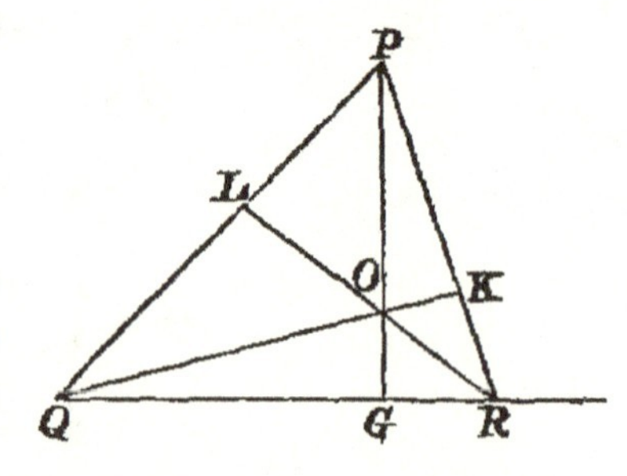

If O be the orthocentre, the triangles LOP, LQR are similar, and
$$LO : LP :: LQ : LR;$$
$$\therefore LO \,.\, LR = LP \,.\, LQ.$$

But, if a rectangular hyperbola pass through P, Q, R, the diameters parallel to LR, PQ are equal: hence O is a point on the curve.

If the angle PRQ is a right angle, the line ROL will be the tangent to the curve at R, so that if a rectangular hyperbola pass through the angular points of a right-angled triangle, the hypotenuse will be parallel to the normal at the right-angle vertex.

144. Prop. X. *If a rectangular hyperbola circumscribe a triangle, the locus of its centre is the nine-point circle of the triangle.*

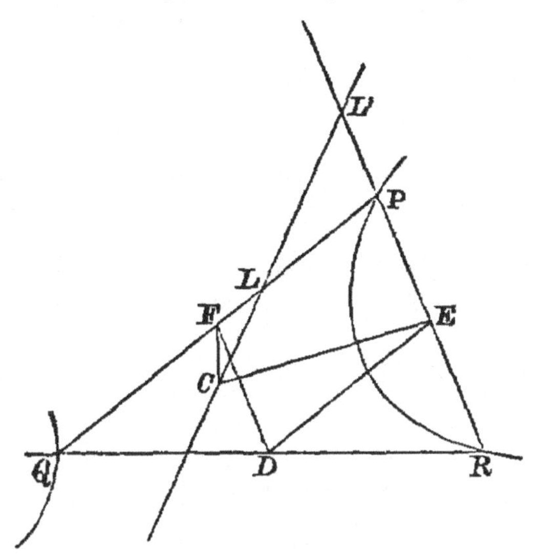

If PQR be the triangle, let L, L' be the points in which an asymptote meets the sides PQ, PR.

Join C, the centre of the hyperbola, with E and F, the middle points of PR and PQ.

Then CF is conjugate to PQ, and CE to PR; therefore the angle

$$\begin{aligned} FCE &= FCL + L'CE = CLF + EL'C \\ &= PLL' + PL'L = FPE \\ &= FDE, \end{aligned}$$

if D be the middle point of QR.

$\therefore$ D, E, F, C are concyclic; that is, C lies on the nine-point circle.

A similar proof is applicable to the case in which the points P, Q, R lie on the same branch of the hyperbola.

EXAMPLES.

1. PCP is a transverse diameter, and QV an ordinate; shew that QV is the tangent at Q to the circle circumscribing the triangle PQp.

2. If the tangent at P meet the asymptotes in L and M, and the normal meet the transverse axis in G, a circle can be drawn through C, L, M, and G, and LGM is a right angle.

3. If AA' be any diameter of a circle, PP' any ordinate to it, then the locus of the intersections of AP, $A'P'$ is a rectangular hyperbola.

4. Given an asymptote and a tangent at a given point, construct the rectangular hyperbola.

5. The points of intersection of an ellipse and a confocal rectangular hyperbola are the extremities of the equi-conjugate diameters of the ellipse.

6. If CP, CD be conjugate semi-diameters, and PN, DM ordinates of any diameter, the triangles PCN, DCM are equal in all respects.

7. The distance of any point from the centre is a geometric mean between its distances from the foci.

8. If P be a point on an equilateral hyperbola, and if the tangent at Q meet CP in T, the circle circumscribing CTQ touches the ordinate QV conjugate to CP.

9. If a circle be described on SS' as diameter, the tangents at the vertices will intersect the asymptotes in the circumference.

10. If two concentric rectangular hyperbolas be described, the axes of one being the asymptotes of the other, they will intersect at right angles.

11. If the tangents at two points Q and Q' meet in T, and if CQ, CQ' meet these tangents in R and R', the points R, T, R', C are concyclic.

12. If from a point Q in the conjugate axis QA be drawn to the vertex, and QR parallel to the transverse axis to meet the curve, $QR = AQ$.

13. Straight lines, passing through a given point, are bounded by two fixed lines at right angles to each other; find the locus of their middle points.

14. Given a point Q and a straight line AB, if a line QCP be drawn cutting AB in C, and P be taken in it, so that, PD being a perpendicular upon AB, CD may be of constant magnitude, the locus of P is a rectangular hyperbola.

15. Every conic passing through the centres of the four circles which touch the sides of a triangle, is a rectangular hyperbola.

16. Ellipses are inscribed in a given parallelogram, shew that their foci lie on a rectangular hyperbola.

17. If two focal chords be parallel to conjugate diameters, the lines joining their extremities intersect on the asymptotes.

18. If P, Q be two points of a rectangular hyperbola, centre O, and QN the perpendicular let fall on the tangent at P, the circle through O, N, and P will pass through the middle point of the chord P, Q.

19. Having given the centre, a tangent, and a point of a rectangular hyperbola, construct the asymptotes.

20. If a right-angled triangle be inscribed in the curve, the normal at the right angle is parallel to the hypotenuse.

21. On opposite sides of any chord of a rectangular hyperbola are described equal segments of circles; shew that the four points, in which the circles, to which these segments belong, again meet the hyperbola, are the angular points of a parallelogram.

22. Two lines of given lengths coincide with and move along two fixed lines, in such a manner that a circle can always be drawn through their extremities; the locus of the centre is a rectangular hyperbola.

23. If a rectangular hyperbola, having its asymptotes coincident with the axes of an ellipse, touch the ellipse, the axis of the hyperbola is a mean proportional between the axes of the ellipse.

24. The tangent at a point P of a rectangular hyperbola meets a diameter QCQ' in T. Shew that CQ and TQ' subtend equal angles at P.

25. If A be any point in a rectangular hyperbola, of which O is the centre, BOC the straight line through O at right angles to OA, D any other point in the curve, and DB, DC parallel to the asymptotes, prove that B, D, A, C are concyclic.

26. The angle subtended by any chord at the centre is the supplement of the angle between the tangents at the ends of the chord.

27. If two rectangular hyperbolas intersect in A, B, C, D; the circles described on AB, CD as diameters intersect each other orthogonally.

28. Prove that the triangle, formed by the tangent at any point and its intercepts on the axes, is similar to the triangle formed by the straight line joining that point with the centre, and the abscissa and ordinate of the point.

29. The angle of inclination of two tangents to a parabola is half a right angle; prove that the locus of their point of intersection is a rectangular hyperbola, having one focus and the corresponding directrix coincident with the focus and directrix of the parabola.

30. P is a point on the curve, and PM, PN are straight lines making equal angles with one of the asymptotes; if MP, NP be produced to meet the curve in P' and Q', then $P'Q'$ passes through the centre.

31. A circle and a rectangular hyperbola intersect in four points and one of their common chords is a diameter of the hyperbola; shew that the other common chord is a diameter of the circle.

32. AB is a chord of a circle and a diameter of a rectangular hyperbola; P any point on the circle; AP, BP, produced if necessary, meet the hyperbola in Q, Q', respectively; the point of intersection of BQ, AQ' will be on the circle.

33. PP' is any diameter, Q any point on the curve, PR, $P'R'$ are drawn at right angles to PQ, $P'Q$ respectively, intersecting the normal at Q in R, R'; prove that QR and QR' are equal.

34. Parallel tangents are drawn to a series of confocal ellipses; prove that the locus of the points of contact is a rectangular hyperbola having one of its asymptotes parallel to the tangents.

35. If tangents, parallel to a given direction, are drawn to a system of circles passing through two fixed points, the points of contact lie on a rectangular hyperbola.

36. If from a point P on the curve chords are equally inclined to the asymptotes, the line joining their other extremities passes through the centre.

37. From the point of intersection of the directrix with one of the asymptotes of a rectangular hyperbola a tangent is drawn to the curve and meets the other asymptote in T; shew that CT is equal to the transverse axis.

38. The normals at the ends of two conjugate diameters intersect on the asymptote, and are parallel to another pair of conjugate diameters.

39. If the base AB of a triangle ABC be fixed, and if the difference of the angles at the base is constant, the locus of the vertex is a rectangular hyperbola.

40. A circle described through the angular points A, B of a given triangle ABC meets AC in D. If BD meet the tangent at A in P, shew that the vertex and orthocentre of the triangle APB lie on fixed rectangular hyperbolas.

41. The locus of the point of intersection of tangents to an ellipse which make equal angles with the transverse and conjugate axes respectively, and are not at right angles, is a rectangular hyperbola whose vertices are the foci of the ellipse.

42. If OT is the tangent at the point O of a rectangular hyperbola, and PQ a chord meeting it at right angles in T, the two bisectors of the angle OCT bisect OP and OQ.

43. With two sides of a square as asymptotes, and the opposite point as focus, a rectangular hyperbola is described; prove that it bisects the other sides.

44. With the focus S of a rectangular hyperbola as centre and radius equal to SC a circle is described, prove that it touches the conjugate hyperbola.

45. If parallel normal chords are drawn to a rectangular hyperbola, the diameter bisecting them is perpendicular to the join of their feet.

46. From the foot of the ordinate PN of a point P of a rectangular hyperbola, tangents NQ, NR are drawn to the circle on AA' as diameter. Prove that PQ passes through A', and PR through A, and that, if QR intersect AA' in M, PM is the tangent at P.

47. Shew that the angle between two tangents to a rectangular hyperbola is equal or supplementary to the angle which their chord of contact subtends at the centre, and that the bisectors of these angles meet on the chord of contact.

48. Through a point P on an equilateral hyperbola two lines are drawn parallel to a pair of conjugate diameters; the one meeting the curve in P, P', and the other meeting the asymptotes in Q, Q'; shew that $PP' = QQ'$.

49. If four points forming a parallelogram be taken on a rectangular hyperbola, then the product of the perpendiculars from any point of the curve on one pair of opposite sides equals the product of the perpendiculars on the other pair of sides.

CHAPTER VI.

THE CYLINDER AND THE CONE.

DEFINITION.

145. If a straight line move so as to pass through the circumference of a given circle, and to be perpendicular to the plane of the circle, it traces out a surface called a *Right Circular Cylinder.* The straight line drawn through the centre of the circle perpendicular to its plane is the *Axis* of the Cylinder.

It is evident that a section of the surface by a plane perpendicular to the axis is a circle, and that a section by any plane parallel to the axis consists of two parallel lines.

PROP. I. *Any section of a cylinder by a plane not parallel or perpendicular to the axis is an ellipse.*

If APA' be the section, let the plane of the paper be the plane through the axis perpendicular to APA'.

Inscribe in the cylinder a sphere touching the cylinder in the circle EF and the plane APA' in the point S.

Let the planes APA', EF intersect in XK, and from any point P of the section draw PK perpendicular to XK.

Draw through P the circular section QP, cutting APA' in PN, so that PN is at right angles to AA' and therefore parallel to XK.

Let the generating line through P meet the circle EF in R; and join SP.

Then PS and PR are tangents to the sphere;

$$\therefore SP = PR = EQ.$$

But $$EQ : NX :: AE : AX$$
$$:: SA : AX,$$
and $$NX = PK,$$
$$\therefore SP : PK :: SA : AX.$$

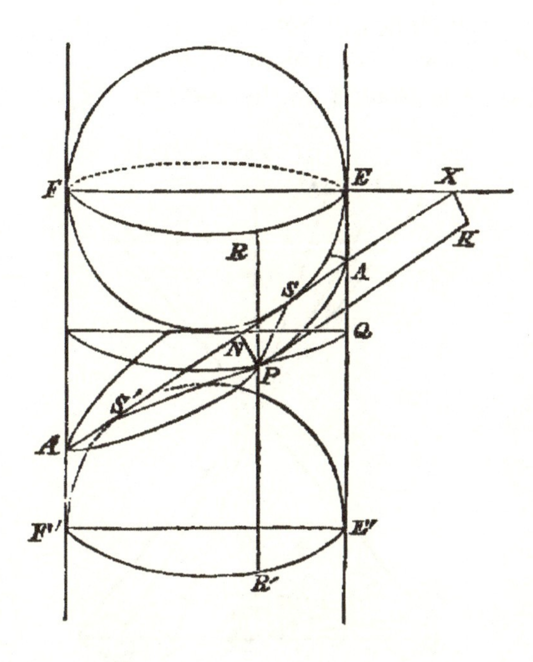

Also, AE being less than AX, SA is less than AX, and the curve APA' is therefore an ellipse, of which S is the focus and XK the directrix.

If another sphere be inscribed in the cylinder touching AA' in S', S' is the other focus, and the corresponding directrix is the intersection of the plane of contact $E'F'$ with APA'.

Producing the generating line RP to meet the circle $E'F'$ in R' we observe that $S'P = PR'$, and therefore

$$\begin{aligned} SP + S'P = RR' &= EE' \\ &= AE + AE' \\ &= AS + AS'; \end{aligned}$$

and $$AS' = AE' = A'F = A'S,$$
$$\therefore SP + S'P = AA'.$$

The transverse axis of the section is AA' and the conjugate, or minor, axis is evidently a diameter of a circular section.

146. DEF. If O be a fixed point in a straight line OE drawn through the centre E of a fixed circle at right angles to the plane of the circle, and if a straight line QOP move so as always to pass through the circumference of the circle, the surface generated by the line QOP is called a *Right Circular Cone.*

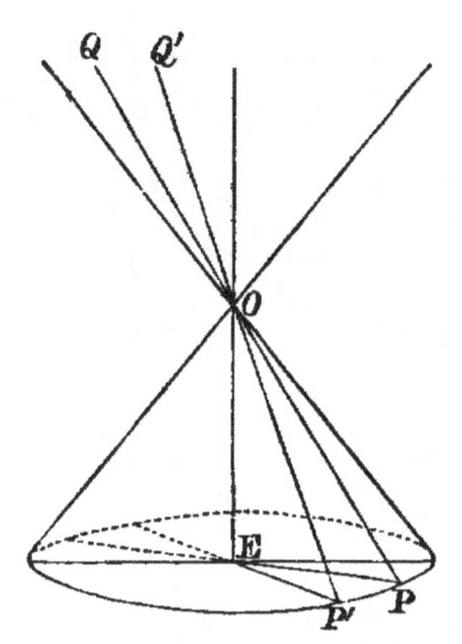

The line OE is called the axis of the cone, the point O is the *vertex*, and the constant angle POE is the semi-vertical angle of the cone.

It is evident that any section by a plane perpendicular to the axis, or parallel to the base of the cone, is a circle; and that any section by a plane through the vertex consists of two straight lines, the angle between which is greatest and equal to the vertical angle when the plane contains the axis.

Any plane containing the axis is called a *Principal Section.*

147. PROP. II. *The section of a cone by a plane, which is not perpendicular to the axis, and does not pass through the vertex, is either an Ellipse, a Parabola, or an Hyperbola.*

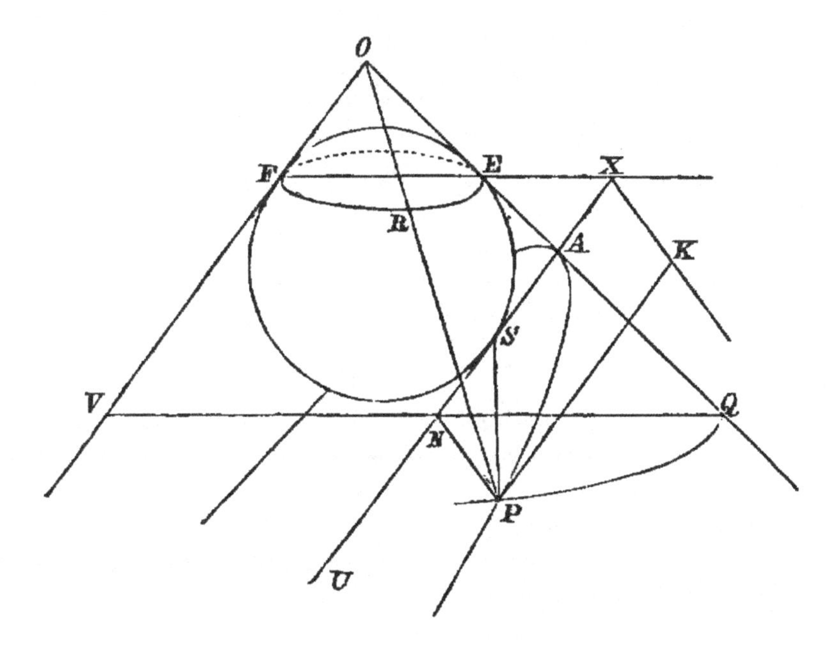

Let UAP be the cutting plane, and let the plane of the paper be that principal section which is perpendicular to the plane UAP; OV, OAQ being the generating lines in the plane of the paper.

Let AU be the intersection of the principal section VOQ by the plane PAU perpendicular to it, and cutting the cone in the curve AP.

Inscribe a sphere in the cone, touching the cone in the circle EF and the plane AP in the point S, and let XK be the intersection of the planes AP, EF. Then XK is perpendicular to the plane of the paper.

Taking any point P in the curve, join OP cutting the circle EF in R, and join SP.

Draw through P the circular section QPV cutting the plane AP in PN which is therefore perpendicular to AN and parallel to XK.

Then, SP and PR being tangents to the sphere,

$$SP = PR = EQ;$$

and
$$EQ : NX :: AE : AX$$
$$:: AS : AX.$$

Also
$$NX = PK;$$
$$\therefore SP : PK :: SA : AX.$$

The curve AP is therefore an Ellipse, Parabola, or Hyperbola, according as SA is less than, equal to, or greater than AX. In any case the point S is a focus and the corresponding directrix is the intersection of the plane of the curve with the plane of contact of the sphere.

148. (1) If AU be parallel to OV, the angle

$$AXE = OFE = OEF = AEX,$$

so that

$$SA = AE = AX;$$

the section is therefore a parabola when the cutting plane is parallel to a generating line, and perpendicular to the principal section which contains the generating line.

(2) Let the line AU meet the curve again in the point A' on the same side of the vertex as the point A.

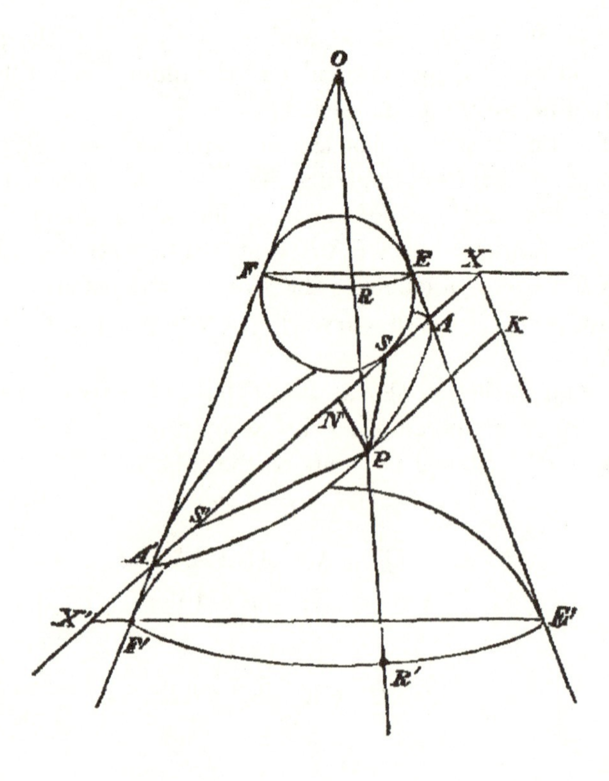

Then the angle

$$AEX = OFX$$
$$> FXA,$$

and therefore $$AE < AX,$$

that is $$SA < AX,$$

and the curve is an ellipse.

In this case another sphere can be inscribed in the cone, touching the cone along the circle $E'F'$ and touching the plane AP in S'.

It may be shewn as before that S' is a focus and that the corresponding directrix is the intersection of the planes $E'F'$, APA'.

(3) Let the line UA produced meet the cone on the other side of the vertex. The section then consists of two separate branches.

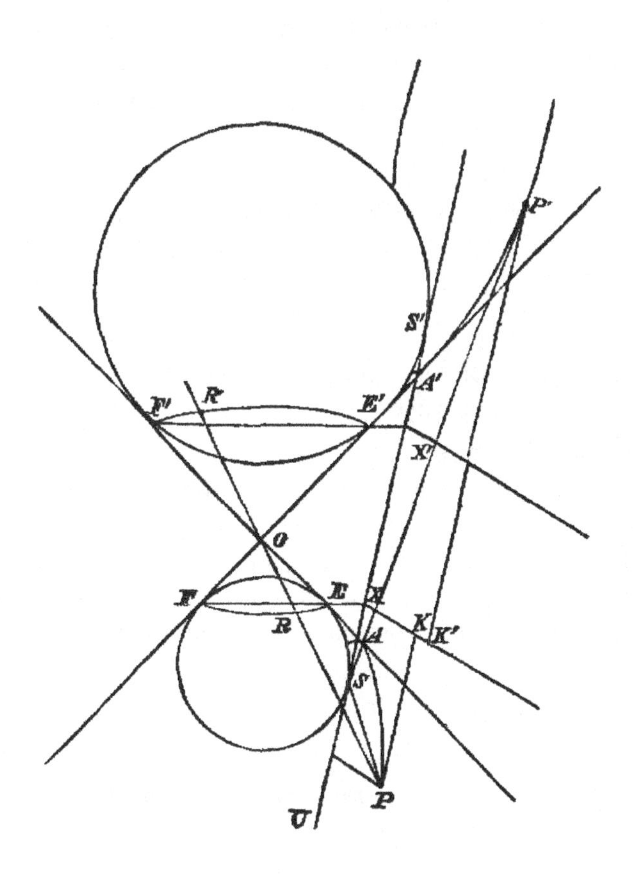

Also the angle $AEX = A'FX$

$< AXF,$

and therefore $AE > AX,$

that is $AS > AX,$

and the curve AP is one branch of an hyperbola, the other branch being the section $A'P'$.

Taking P' in the other branch the proof is the same as before that

$$SP' : P'K' :: SA : AX.$$

In this case a sphere can be inscribed in the other branch of the cone, touching the cone along the circle $E'F'$, and the plane $UA'P'$ in S', and it can be shewn that S' is the other focus of the hyperbola, and that the directrix is the intersection of the cutting plane with the plane of contact $E'F'$.

Hence the section of a cone by a plane cutting in AU the principal section VOQ perpendicular to it is an Ellipse, Parabola, or Hyperbola, according as the angle EAX is greater than, equal to, or less than, the vertical angle of the cone.

Further, it is obvious that, if any plane be drawn parallel to the plane AP, the ratio of AE to AX is always the same; hence it follows that all parallel sections have the same eccentricity.

149. This method of determining the focus and directrix was published by Mr Pierce Morton, of Trinity College, in the first volume of the Cambridge *Philosophical Transactions*.

The method was very nearly obtained by Hamilton, who gave the following construction.

First finding the vertex and focus, A and S, take AE along the generating line equal to AS, and draw the circular section through E; the directrix will be the line of intersection of the plane of the circle with the given plane of section.

Hamilton also demonstrated the equality of SP and PR.

150. Prop. III. *To prove that, in the case of an elliptic section,*

$$SP + S'P = AA'.$$

Taking the 2nd figure,

$$SP = PR \text{ and } S'P = PR';$$
$$\therefore SP + S'P = RR' = EE'$$
$$= AE + AE'$$
$$= AS + AS'.$$

But
$$A'S' = A'F' = FF' - A'F$$
$$= EE' - A'S,$$

also
$$A'S' + SS' = A'S;$$
$$\therefore 2A'S' + SS' = EE'.$$

Similarly
$$2AS + SS' = EE';$$
$$\therefore A'S' = AS,$$

and
$$AS' = A'S.$$

Hence
$$SP + S'P = AA',$$

and the transverse, or major axis $= EE'$.

In a similar manner it can be shewn that in an hyperbolic section

$$S'P - SP = AA'.$$

151. Prop. IV. *To shew that, in a parabolic section,*

$$PN^2 = 4AS \,.\, AN.$$

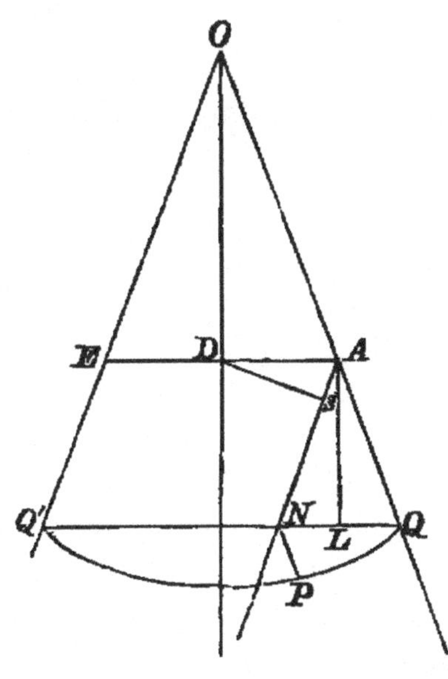

Let A be the vertex of the section, and let ADE be the diameter of the circular section through A. From D let fall DS perpendicular to AN;

then
$$PN^2 = QN \,.\, NQ'$$
$$= QN \,.\, AE$$
$$= 4NL \,.\, AD,$$

if AL be perpendicular to NQ.

But the triangles ANL, ADS being similar,
$$NL : AN :: AS : AD;$$
$$\therefore NL \,.\, AD = AN \,.\, AS,$$

and
$$PN^2 = 4AS \,.\, AN.$$

152. Prop. V. *To shew that, in an elliptic section, PN^2 is to $AN \,.\, NA'$ in a constant ratio.*

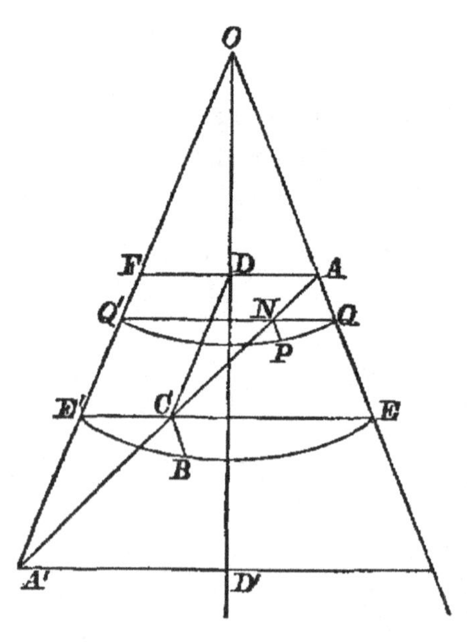

Draw through P the circular section QPQ', bisect AA' in C, and draw through C the circular section EBE'.

Then
$$QN : AN :: CE : AC,$$
and
$$NQ' : NA' :: CE' : A'C;$$
$$\therefore QN \,.\, NQ' : AN \,.\, NA' :: EC \,.\, CE' : AC^2,$$

or
$$PN^2 : AN \,.\, NA' :: EC \,.\, CE' : AC^2;$$

and, the transverse axis being AA', the square of the semi-minor axis $= BC^2 = EC.CE'$. Again, if ADF be perpendicular to the axis, $AD = DF$, and, AC being equal to CA', CD is parallel to $A'F$, and therefore
$$CE' = FD = AD.$$

Similarly, $CE = A'D'$, the perpendicular from A' on the axis;
$$\therefore BC^2 = AD \,.\, A'D',$$

that is, *the semi-minor axis is a mean proportional between the perpendiculars from the vertices on the axis of the cone.*

COR. If H, H' are the centres of the focal spheres, the angles HAH', $HA'H'$ are right angles, so that H, A, H', A' are concyclic.

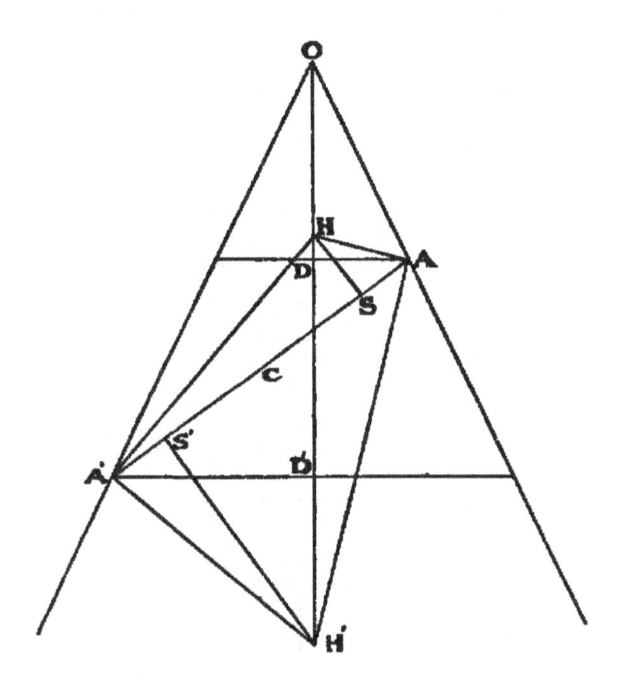

It follows that the triangles ASH, $A'H'D'$ are similar, as are also the triangles $A'S'H'$, AHD, so that

$$SH : A'D' :: AH : A'H' :: AD : S'H';$$

and $$SH \,.\, S'H' = AD \,.\, A'D' = BC^2;$$

$\therefore$ *the semi-minor axis is a mean proportional between the radii of the focal spheres.*

The fact that H, A, H', A' are concyclic also shews that the sphere of which HH' is a diameter intersects the plane of the ellipse in its auxiliary circle.

153. In exactly the same manner it can be shewn that, for an hyperbolic section,

$$PN^2 : AN \,.\, NA' :: CE \,.\, CE' : AC^2,$$

and that $$CE = AD, \text{and } CE' = A'D'.$$

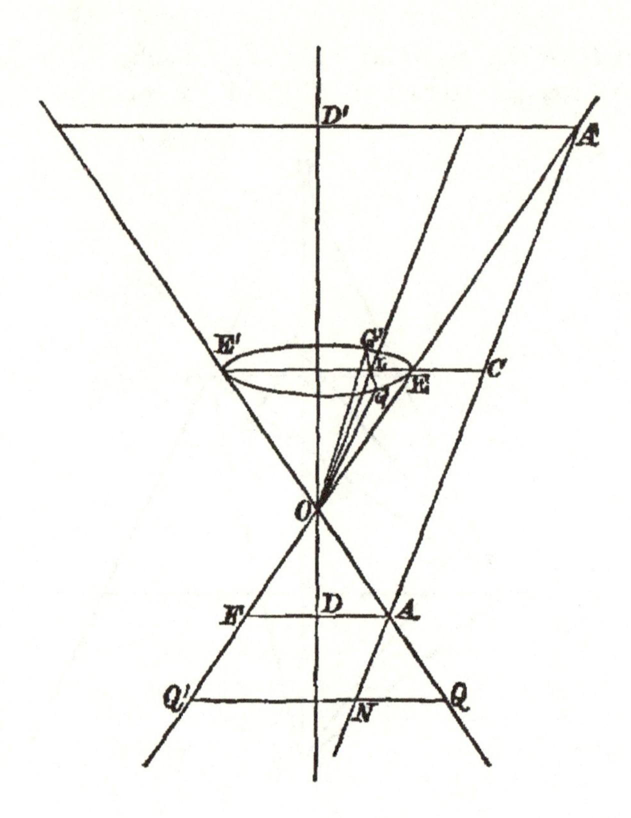

Also, as in the case of the ellipse, BC is a mean proportional between AD and $A'D'$, and is also a mean proportional between the radii of the focal spheres.

154. PROP. VI. *The two straight lines in which a cone is intersected by a plane through the vertex parallel to an hyperbolic section are parallel to the asymptotes of the hyperbola.*

Taking the preceding figure, let the parallel plane cut the cone in the lines OG, OG', and the circular section through C in the line GLG', which will be perpendicular to the plane of the paper, and therefore perpendicular to EE' and to OL.

Hence $$GL^2 = EL \,.\, E'L.$$

But $$EL : EC :: OL : A'C,$$
and $$E'L : E'C :: OL : AC;$$
$$\therefore GL^2 : EC . E'C :: OL^2 : AC^2,$$
or $$GL : OL :: BC : AC;$$
therefore, (Art. 102), OG and OG' are parallel to the asymptotes of the hyperbola.

Hence, for all parallel hyperbolic sections, the asymptotes are parallel to each other.

If the hyperbola be rectangular, the angle GOG' is a right angle; but this is evidently not possible if the vertical angle of the cone be less than a right angle.

When the vertical angle of the cone is not less than a right angle, and when GOG' is a right angle, LOG is half a right angle, and therefore
$$OL = LG,$$
and $$2 . OL^2 = OG^2 = OE^2,$$
and the length OL is easily constructed.

Hence, placing OL, and drawing the plane GOG' perpendicular to the principal section through OL, any section by a plane parallel to GOG' is a rectangular hyperbola.

It will be observed that the eccentricity of the section is greatest when its plane is parallel to the axis of the cone.

155. PROP. VII. *The sphere which passes through the circles of contact of the focal spheres with the surface of the cone intersects the plane of the section in its director circle.*

Let Q, Q' be the points in which the straight line AA' is intersected by the sphere which passes through the circles EF and $E'F'$.

Then the sphere intersects the plane of the ellipse in the circle of which QQ' is the diameter.

Also $$CQ^2 - CA^2 = AQ . AQ' = AE . AE'$$
$$= AS . AS' = BC^2;$$
$$\therefore CQ^2 = AC^2 + BC^2,$$
so that CQ is the radius of the director circle.

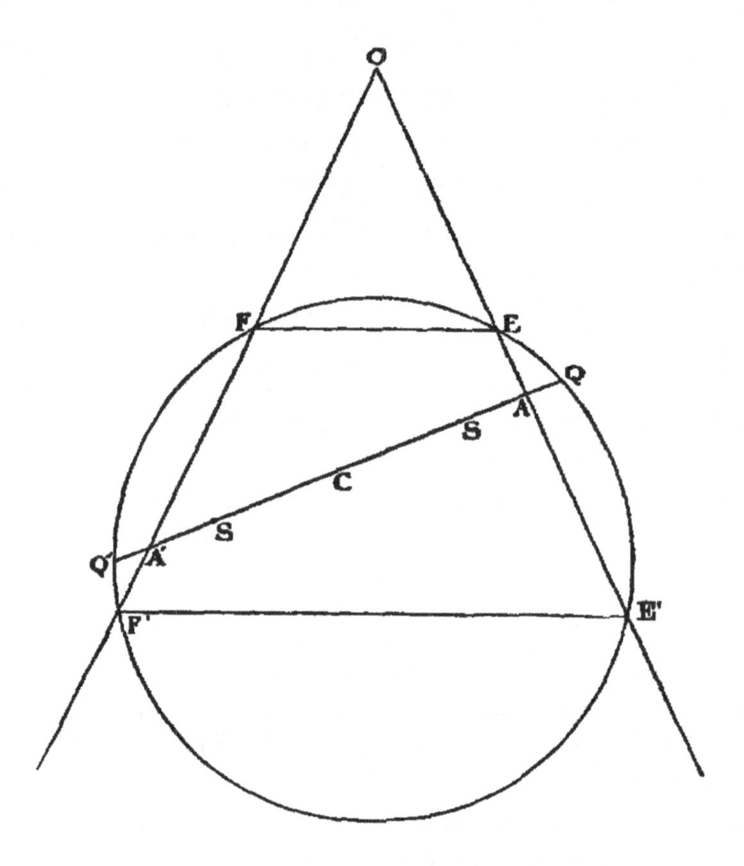

Changing the figure the proof is exactly the same for the hyperbola.

156. PROP. VIII. *If two straight lines be drawn through any point, parallel to two fixed lines, and intersecting a given cone, the ratio of the rectangles formed by the segments of the lines will be independent of the position of the point.*

Thus, if through E, the lines EPQ, $EP'Q'$ be drawn, parallel to two given lines, and cutting the cone in the points P, Q and P', Q', the ratio of $EP \cdot EQ$ to $EP' \cdot EQ'$ is constant.

Through O draw OK parallel to the given line to which EPQ is parallel, and let the plane through OK, EPQ, which contains the generating lines OP, OQ, meet the circular section through E in R and S, and the plane base in the straight line DFK, cutting the circular base in D and F.

Then DFK and ERS being sections of parallel planes by a plane are parallel to each other.

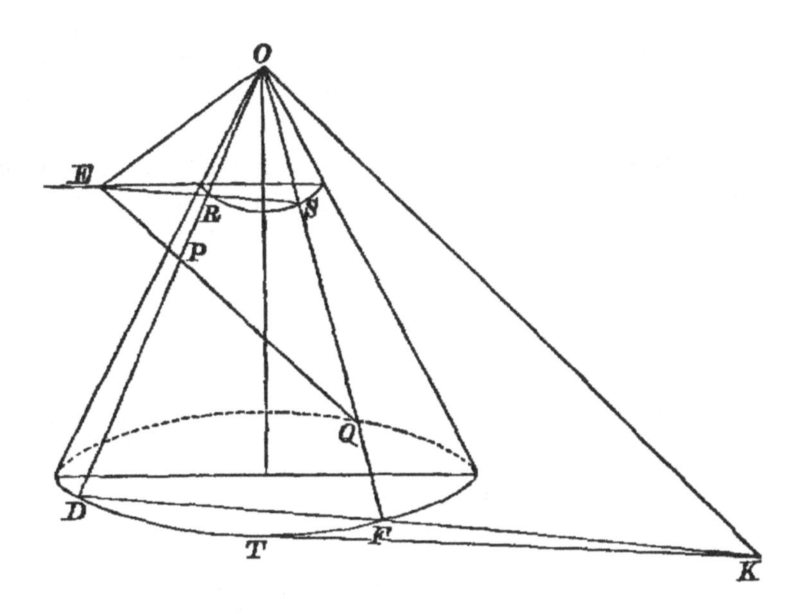

Also, EPQ is parallel to OK;

Therefore ERP, ODK are similar triangles, as are also ESQ, OFK;

$$\therefore EP : ER :: OK : DK,$$

and

$$EQ : ES :: OK : FK;$$

$$\therefore EP \,.\, EQ : ER \,.\, ES :: OK^2 : DK \,.\, FK$$

$$:: OK^2 : KT^2,$$

if KT be the tangent to the circular base from K.

If a similar construction be made for $EP'Q'$, we shall have

$$EP' \,.\, EQ' : ER' \,.\, ES' :: OK'^2 : K'T'^2.$$

But

$$ER \,.\, ES = ER' \,.\, ES';$$

therefore the rectangles $EP \,.\, EQ$ and $EP' \,.\, EQ'$ are each in a constant ratio to the same rectangle, and are therefore in a constant ratio to each other.

Since the plane through EPQ, $EP'Q'$ cuts the cone in an ellipse, parabola, or hyperbola, this theorem includes as particular cases those of Arts. 51, 58, 82, 92, 96, 124 and 134.

The proof is the same if the point P be within the cone, or if one or both of the lines meet opposite branches of the cone.

If the chords be drawn through the centre of the section PEP', the rectangles become the squares of the semi-diameters.

Hence the parallel diameters of all parallel sections of a cone are proportional to each other.

If the lines move until they become tangents the rectangles then become the squares of the tangents; therefore if a series of points be so taken that the tangents from them are parallel to given lines, these tangents are always in the same proportion. The locus of the point E will be the line of intersection of two fixed planes touching the cone, that is, a fixed line through the vertex.

EXAMPLES.

1. Shew how to cut from a cylinder an ellipse of given eccentricity.

2. What is the locus of the foci of all sections of a cylinder of a given eccentricity?

3. Shew how to cut from a cone an ellipse of given eccentricity.

4. Prove that all sections of a cone by parallel planes are conics of the same eccentricity.

5. What is the locus of the foci of the sections made by planes inclined to the axis at the same angle?

6. Find the least angle of a cone from which it is possible to cut an hyperbola, whose eccentricity shall be the ratio of two to one.

7. The centre of a spherical ball is moveable in a vertical plane which is equidistant from two candles of the same height on a table; find its locus when the two shadows on the ceiling are always just in contact.

8. Through a given point draw a plane cutting a given cone in a section which has the given point for a focus.

9. If the vertical angle of a cone, vertex O, be a right angle, P any point of a parabolic section, and PN perpendicular to the axis of the parabola,

$$OP = 2AS + AN,$$

A being the vertex and S the focus.

10. Prove that the directrices of all parabolic sections of a cone lie in the tangent planes of a cone having the same axis.

11. If the curve formed by the intersection of any plane with a cone be projected upon a plane perpendicular to the axis; prove that the curve of projection will be a conic section having its focus at the point in which the axis meets the plane of projection.

12. Prove that the latera recta of parabolic sections of a right circular cone are proportional to the distances of their vertices from the vertex of the cone.

13. The shadow of a ball is cast by a candle on an inclined plane in contact with the ball; prove that as the candle burns down, the locus of the centre of the shadow will be a straight line.

14. The vertex of a right cone which contains a given ellipse lies on a certain hyperbola, and the axis of the cone will be a tangent to the hyperbola.

15. Find the locus of the vertices of the right circular cones which can be drawn so as to pass through a given fixed hyperbola, and prove that the axis of the cone is always tangential to the locus.

16. An ellipse and an hyperbola are so situated that the vertices of each curve are the foci of the other, and the curves are in planes at right angles to each other. If P be a point on the ellipse, and O a point on the hyperbola, S the vertex, and A the interior focus of that branch of the hyperbola, then

$$AS + OP = AO + SP.$$

17. The latus rectum of any plane section of a given cone is proportional to the perpendicular from the vertex on the plane.

18. If a sphere is described about the vertex of a right cone as centre, the latera recta of all sections made by tangent planes to the sphere are equal.

19. Different elliptic sections of a right cone are taken such that their minor axes are equal; shew that the locus of their centres is the surface formed by the revolution of an hyperbola about the axis of the cone.

20. If two cones be described touching the same two spheres, the eccentricities of the two sections of them made by the same plane bear to one another a ratio constant for all positions of the plane.

21. If elliptic sections of a cone be made such that the volume between the vertex and the section is always the same, the minor axis will be always of the same length.

22. The vertex of a cone and the centre of a sphere inscribed within it are given in position: a plane section of the cone, at right angles to any generating line of the cone, touches the sphere: prove that the locus of the point of contact is a surface generated by the revolution of a circle, which touches the axis of the cone at the centre of the sphere.

23. Given a right cone and a point within it, there are two sections which have this point for focus; and the planes of these sections make equal angles with the straight line joining the given point and the vertex of the cone.

24. Prove that the centres of all plane sections of a cone, for which the distance between the foci is the same, lie on the surface of a right circular cylinder.

CHAPTER VII.

The Similarity of Conics, the Areas of Conics, and the Curvatures of Conics.

SIMILAR CONICS.

157. Def. *Conics which have the same eccentricity are said to be similar to each other.*

This definition is justified by the consideration that the character of the conic depends on its eccentricity alone, while the dimensions of all parts of the conic are entirely determined by the distance of the focus from the directrix.

Hence, according to this definition, all parabolas are similar curves.

Prop. I. *If radii be drawn from the vertices of two parabolas making equal angles with the axis, these radii are always in the same proportion.*

Let AP, ap be the radii, PN and pn the ordinates, the angles PAN, pan, being equal.

Then $$AP^2 : ap^2 :: PN^2 : pn^2 :: AS \,.\, AN : as \,.\, an.$$

But $$AP : ap :: AN : an;$$

$$\therefore AP : ap :: AS : as.$$

It can also be shewn that focal radii making equal angles with the axes are always in the same proportion.

158. Prop. II. *If two ellipses be similar their axes are in the same proportion, and any other diameters, making equal angles with the respective axes, are in the proportion of the axes.*

Let CA, CB be the semi-axes of one ellipse, ca, cb of the other, and CP, cp two radii such that the angle $PCA = pca$.

Then, since the eccentricities are the same, we have, if S, s be foci,

$$AC : SC :: ac : sc;$$

$$\therefore AC^2 : AC^2 - SC^2 :: ac^2 : ac^2 - sc^2,$$

or
$$AC^2 : BC^2 :: ac^2 : bc^2.$$

Hence it follows, if PN, pn be ordinates, that

$$PN^2 : AC^2 - CN^2 :: pn^2 : ac^2 - cn^2;$$

but, by similar triangles,

$$PN : pn :: CN : cn,$$

therefore
$$CN^2 : AC^2 - CN^2 :: cn^2 : ac^2 - cn^2;$$

and
$$CN^2 : AC^2 :: cn^2 : ac^2.$$

Hence
$$CP : cp :: CN : cn$$
$$:: AC : ac.$$

So also lines drawn similarly from the foci, or any other corresponding points of the two figures, will be in the ratio of the transverse axes.

Exactly the same demonstration is applicable to the hyperbola, but in this case, if the ratio of SC to AC in two hyperbolas be the same, it follows from Art. (102) that the angle between the asymptotes is the same in both curves.

In the case of hyperbolas we have thus a very simple test of similarity.

The Areas bounded by Conics.

159. Prop. III. *If AB, AC be two tangents to a parabola, the area between the curve and the chord BC is two-thirds of the triangle ABC.*

Draw the tangent DPE parallel to BC; then

$$AP = PN,$$

and
$$BC = 2 . DE;$$

therefore the triangle
$$BPC = 2ADE.$$

Again, draw the diameter DQM meeting BP in M.

By the same reasoning, FQG being the tangent parallel to BP, the triangle $PQB = 2FDG$.

Through F draw the diameter FRL, meeting PQ in L, and let this process be continued indefinitely.

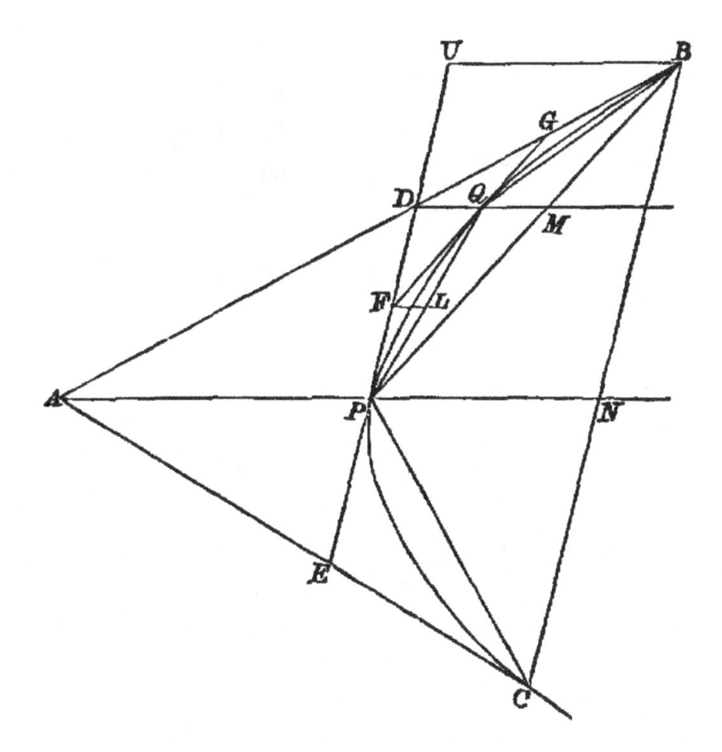

Then the sum of the triangles within the parabola is double the sum of the triangles without it.

But, since the triangle BPC is half ABC, it is greater than half the parabolic area $BQPC$;

Therefore (Euclid, Bk. XII.) the difference between the parabolic area and the sum of the triangles can be made ultimately less than any assignable quantity;

And, the same being true of the outer triangles, it follows that the area between the curve and BC is double of the area between the curve and AB, AC, and is therefore two-thirds of the triangle ABC.

COR. Since PN bisects every chord parallel to BC, it bisects the parabolic area BPC; therefore, completing the parallelogram $PNBU$, the parabolic area BPN is two-thirds of the parallelogram UN.

160. PROP. IV. *The area of an ellipse is to the area of the auxiliary circle in the ratio of the conjugate to the transverse axis.*

Draw a series of ordinates, QPN, $Q'P'N'$, ... near each other, and draw PR, QR' parallel to AC.

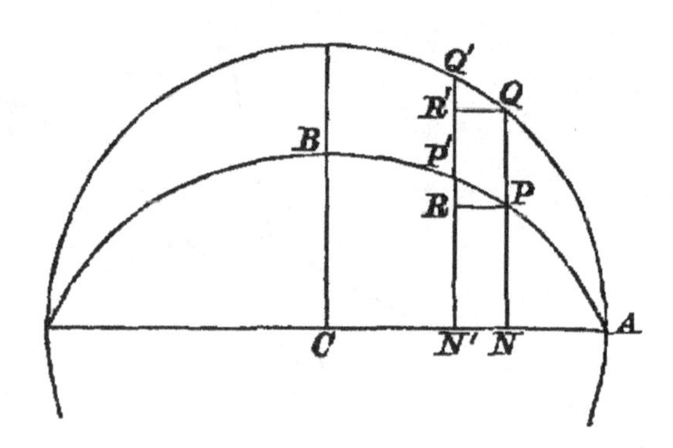

Then, since

$$PN : QN :: BC : AC,$$

the area $$PN' : QN' :: BC : AC,$$

and, this being true for all such areas, the sum of the parallelograms PN' is to the sum of the parallelograms QN' as BC to AC.

But, if the number be increased indefinitely, the sums of these parallelograms ultimately approximate to the areas of the ellipse and circle.

Hence the ellipse is to the circle in the ratio of BC to AC.

The student will find in Newton's 2nd and 3rd Lemmas (*Principia*, Section I.) a formal proof of what we have here assumed as sufficiently obvious, that the sum of the parallelograms PN is ultimately equal to the area of the ellipse.

161. PROP. V. *If P, Q be two points of an hyperbola, and if PL, QM parallel to one asymptote meet the other in L and M, the hyperbolic sector CPQ is equal to the hyperbolic trapezium $PLMQ$.*

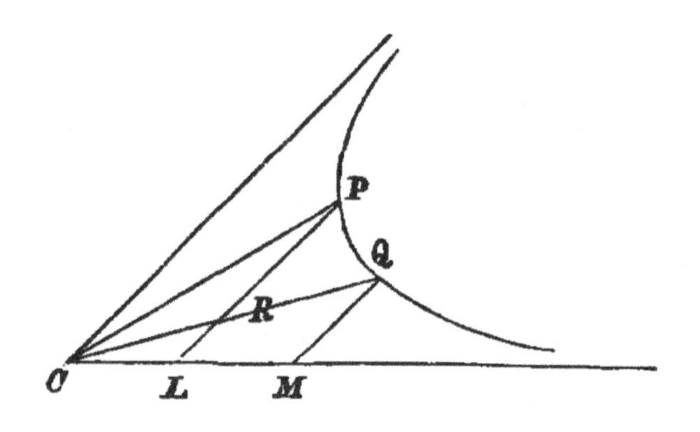

For the triangles CPL, CQM are equal, and, if PL meet CQ in R, it follows that the triangle CPR = the trapezium $LRQM$; hence, adding to each the area RPQ, the theorem is proved.

162. PROP. VI. *If points L, M, N, K be taken in an asymptote of an hyperbola, such that*

$$CL : CM :: CN : CK,$$

and if LP, MQ, NR, KS, parallel to the asymptote, meet the curve in P, Q, R, S, the hyperbolic areas CPQ, CRS will be equal.

Let QR and PS produced meet the asymptotes in F, F', G, G';

then $$RF = QF' \text{ and } SG = PG' \text{ (Art. 121)},$$

$$\therefore NF = CM \text{ and } KG = CL.$$

Hence $$NF : KG :: CM : CL$$
$$:: CK : CN$$
$$:: RN : SK,$$

and therefore SP is parallel to QR.

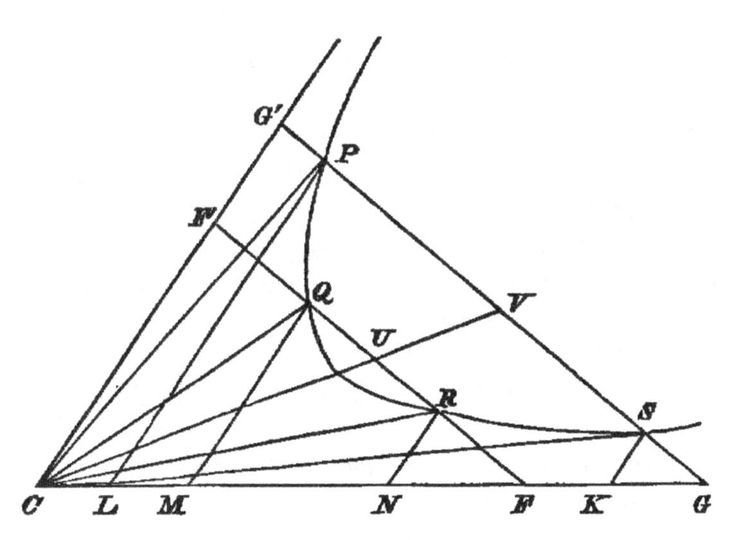

The diameter CUV conjugate to PS bisects all chords parallel to PS, and therefore bisects the area $PQRS$;

also the triangle $$CPV = CSV,$$

and $$CQU = CUR;$$

therefore, taking from CPV and CSV the equal triangles CQU, CRU, and the equal areas $PQUV$, $SRUV$, the remaining areas, which are the hyperbolic sectors CPQ, CRS, are equal.

Cor. Hence if a series of points, L, M, N, ... be taken such that CL, CM, CN, CK, ... are in continued proportion, it follows that the hyperbolic sectors CPQ, CQR, CRS, &c. will be all equal.

It will be noticed in this case that the tangent at Q will be parallel to PR, the tangent at R parallel to QS, and so also for the rest.

The Curvature of Conics.

163. Def. If a circle touch a conic at a point P, and pass through another point Q of the conic, and if the point Q move near to, and ultimately coincide with P, the circle in its ultimate condition is called the circle of curvature at P.

Prop. VII. *The chord of intersection of a conic with the circle of curvature at any point is inclined to the axis at the same angle as the tangent at the point.*

It has been shewn that, if a circle intersect a conic in four points P, Q, R, V, the chords PQ, RV are equally inclined to the axis.

Let P and Q coincide with each other; then the tangent at P and the chord RV are equally inclined to the axis.

Let the point V now approach to and coincide with P; the circle becomes the circle of curvature at P, and the chord VR becomes PR the chord of intersection.

Hence PR and the tangent at P are equally inclined to the axis.

164. Prop. VIII. *If the tangent at any point P of a parabola meet the axis in T, and if the circle of curvature at P meet the curve in Q,*

$$PQ = 4 \, . \, PT.$$

Draw the ordinate PNP'; then taking the figure of the next article, TP' is the tangent at P',

and the angle $$P'TF = PTF = PFT;$$

therefore PQ is parallel to TP', and is bisected by the diameter $P'E$.

Hence $$PQ = 2 \, . \, PE = 4P'T = 4PT.$$

165. PROP. IX. *To find the chord of curvature through the focus and the diameter of curvature at any point of a parabola.*

Let the circle meet PS produced in V, and the normal PG produced in O.

The angle
$$PFS = PTS = SPT$$
$$= PQV,$$
since PT is a tangent to the circle.

Therefore QV is parallel to the axis,

and
$$PV : SP :: PQ : PF.$$

Hence
$$PV = 4 . SP.$$

Again, the angle $POQ = PVQ = PSN$;

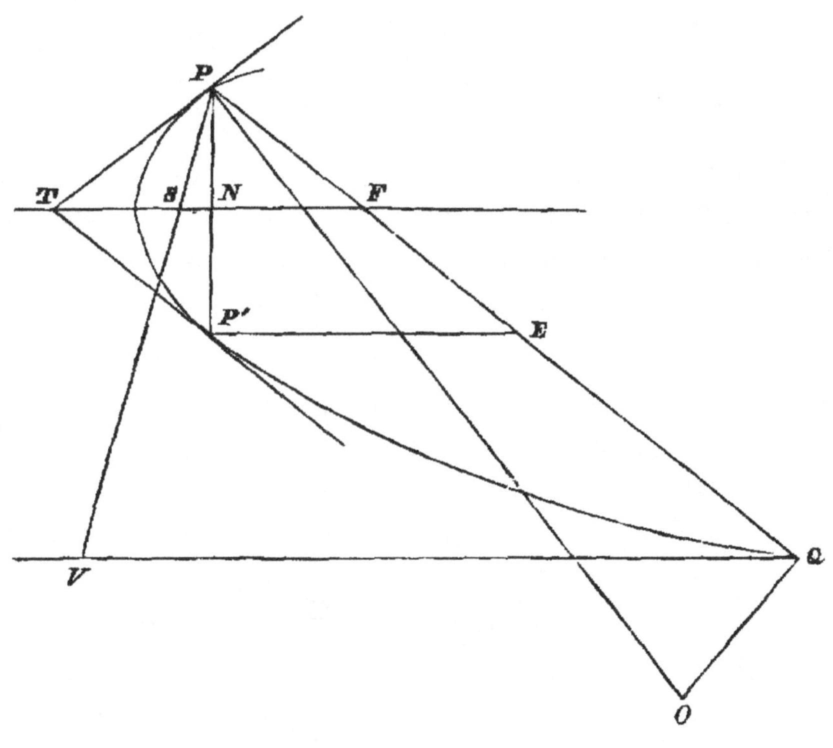

$$\therefore PO : PQ :: SP : PN,$$
or
$$PO : SP :: 4PT : PN$$
$$:: 4SP : SY,$$
if SY be perpendicular to PT.

Cor. 1. Since the normal bisects the angle between SP and the diameter through P, it follows that the chord of curvature parallel to the axis is $4SP$.

Cor. 2. The diameter of curvature, PO, may also be expressed as follows:

Let GL be the perpendicular from G on SP;
then $PL =$ the semi-latus rectum $= 2AS$.

Also PVO being a right angle,

$$PO : PG :: PV : PL$$
$$:: 4SP : PL$$
$$:: 4SP \,.\, PL : PL^2;$$

but
$$4SP \,.\, PL = 8SP \,.\, AS = 8SY^2 = 2PG^2;$$
$$\therefore PO : PG :: 2PG^2 : PL^2.$$

166. Prop. X. *If the chord of intersection, PQ, of an ellipse, or hyperbola, with the circle of curvature at P, meet CD, the semi-diameter conjugate to CP, in K,*

$$PQ \,.\, PK = 2CD^2.$$

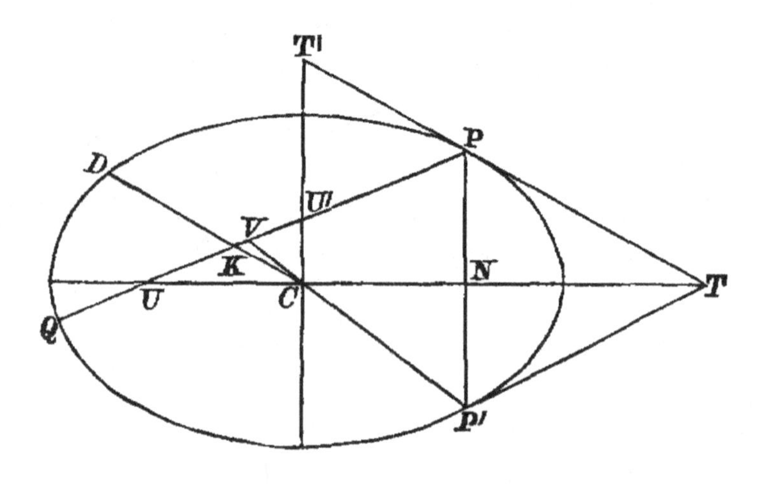

Drawing the ordinate PNP', the tangent at P' is parallel to PQ, as in the parabola, and PQ is therefore bisected in V, by the diameter CP'.

Let PQ meet the axes in U and U'; then, $U'C$ being parallel to PP',

$$PV : PU' :: VP' : CP'$$
$$:: UT : CT,$$

since PU, $P'T$ are parallel.

Also

$$UT : CT :: PU : PK;$$
$$\therefore PV : PU' :: PU : PK.$$

Hence

$$PV \,.\, PK = PU \,.\, PU'$$
$$= PT \,.\, PT' = CD^2,$$

observing that $PU = PT$, and $PU' = PT'$, by the theorem of Art. 163,

and $\qquad \therefore PQ \,.\, PK = 2CD^2.$

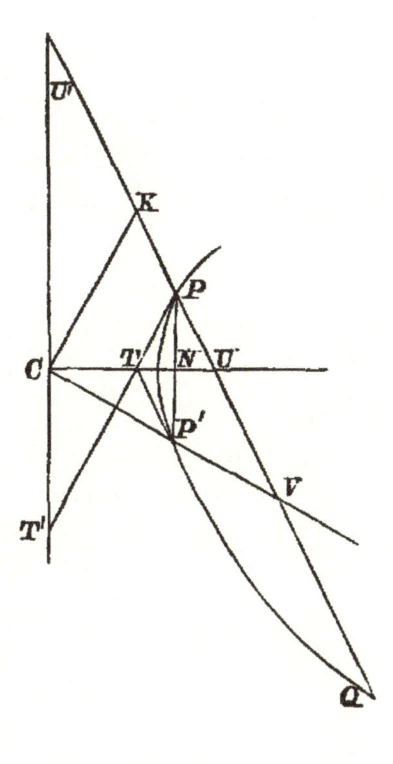

167. Prop. XI. *If the chord of curvature PQ', of an ellipse or hyperbola in any direction, meet CD in K',*

$$PQ' \,.\, PK' = 2CD^2.$$

Let PO be the diameter of curvature meeting CD in F: then PQO, $PQ'O$ are right angles, and a circle can be drawn through $Q'K'FO$;

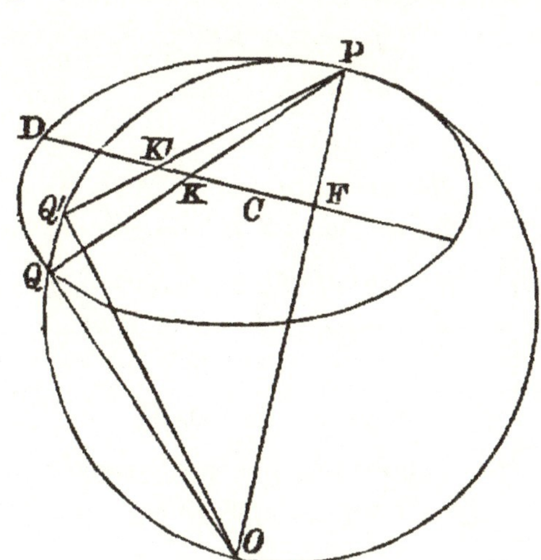

$$\therefore PQ' \,.\, PK' = PF \,.\, PO$$
$$= PK \,.\, PQ = 2 \,.\, CD^2.$$

COR. 1. Hence PO being the diameter of curvature,

$$PF \,.\, PO = 2 \,.\, CD^2.$$

COR. 2. If PQ' pass through the focus,

$$PK' = AC,$$

and
$$PQ' \,.\, AC = 2 \,.\, CD^2.$$

COR. 3. If PQ' pass through the centre,

$$PQ' \,.\, CP = 2 \,.\, CD^2.$$

168. We can also express the diameter of curvature as follows:

PG being the normal, let GL be perpendicular to SP, and let PR be the chord of curvature through S.

Then GL is parallel to OR,

and
$$PO : PG :: PR : PL$$
$$:: PR \,.\, PL : PL^2.$$

But
$$PR \,.\, AC = 2 \,.\, CD^2;$$
$$\therefore PR : AC :: 2 \,.\, CD^2 : AC^2$$
$$:: 2 \,.\, PG^2 : BC^2,$$

and
$$PR \,.\, PL : AC \,.\, PL :: 2 \,.\, PG^2 : BC^2.$$

But, PL being equal to the semi-latus rectum,

$$PL \,.\, AC = BC^2;$$
$$\therefore PR \,.\, PL = 2 \,.\, PG^2,$$

and
$$PO : PG :: 2PG^2 : PL^2.$$

Hence, in any conic, the radius of curvature at any point is to the normal at the point as the square of the normal to the square of the semi-latus rectum.

169. PROP. XII. *The chord of curvature through the focus at any point is equal to the focal chord parallel to the tangent at the point.*

Since
$$PQ' \,.\, AC = 2CD^2,$$

it follows that
$$PQ' \,.\, AA' = DD'^2.$$

But, if pp' is the focal chord parallel to the tangent at P,

$$pp' \,.\, AA' = DD'^2 \text{ (Art. 81)},$$
$$\therefore PQ' = pp'.$$

EXAMPLES.

1. The radius of curvature at the end of the latus rectum of a parabola is equal to twice the normal.

2. The circle of curvature at the end of the latus rectum intersects the parabola on the normal at that point.

3. If PV is the chord of curvature through the focus, what is the locus of the point V?

4. An ellipse and a parabola, whose axes are parallel, have the same curvature at a point P and cut one another in Q; if the tangent at P meets the axis of the parabola in T prove that $PQ = 4 \,.\, PT$.

5. In a rectangular hyperbola, the radius of curvature at P varies as CP^3.

6. If P be a point of an ellipse equidistant from the axis minor and one of the directrices, the circle of curvature at P will pass through one of the foci.

7. If the normal at a point P of a parabola meet the directrix in L, the radius of curvature at P is equal to $2 \,.\, PL$.

8. The normal at any point P of a rectangular hyperbola meets the curve again in Q; shew that PQ is equal to the diameter of curvature at P.

9. In the rectangular hyperbola, if CP be produced to Q, so that $PQ = CP$, and QO be drawn perpendicular to CQ to intersect the normal in O, O is the centre of curvature at P.

10. At any point of an ellipse the chord of curvature PV through the centre is to the focal chord pp', parallel to the tangent, as the major axis is to the diameter through the point.

11. If the common tangent of an ellipse and its circle of curvature at P be bisected by their common chord, prove that

$$CD^2 = AC \,.\, BC.$$

12. The tangent at a point P of an ellipse whose centre is C meets the axes in T and t; if CP produced meet in L the circle described about the triangle TCt, shew that PL is half the chord of curvature at P in the direction of C, and that the rectangle contained by CP, CL, is constant.

13. If P be a point on a conic, Q a point near it, and if QE, perpendicular to PQ, meet the normal at P in E, then ultimately when Q coincides with P, PE is the diameter of curvature at P.

14. If a tangent be drawn from any point of a parabola to the circle of curvature at the vertex, the length of the tangent will be equal to the abscissa of the point measured along the axis.

15. The circle of curvature at a point where the conjugate diameters are equal, meets the ellipse again at the extremity of the diameter.

16. The chord of curvature at P perpendicular to the major axis is to PM, the ordinate at P, $:: 2 . CD^2 : BC^2$.

17. Prove that there is a point P on an ellipse such that if the normal at P meet the ellipse in Q, PQ is a chord of the circle of curvature at P, and find its position.

18. The chord of curvature at a point P of a rectangular hyperbola, perpendicular to an asymptote, is to $CD :: CD : 2 . PN$, where PN is the distance of P from the asymptote.

19. If G be the foot of the normal at a point P of an ellipse, and GK, perpendicular to PG, meet CP in K, then KE, parallel to the axis minor, will meet PG in the centre of curvature at P.

20. The chord of curvature through the vertex at a point of a parabola is to $4PY :: PY : AP$.

21. Prove that the locus of the middle points of the common chords of a given parabola and its circles of curvature is a parabola, and that the envelope of the chords is also a parabola.

22. The circles of curvature at the extremities P, D of two conjugate diameters of an ellipse meet the ellipse again in Q, R, respectively, shew that PR is parallel to DQ.

23. The tangent at any point P in an ellipse, of which S and H are the foci, meets the axis major in T, and TQR bisects HP in Q and meets SP in R; prove that PR is one-fourth of the chord of curvature at P through S.

24. An ellipse, a parabola, and an hyperbola, have the same vertex and the same focus; shew that the curvature, at the vertex, of the parabola is greater than that of the hyperbola, and less than that of the ellipse.

25. The circle of curvature at a point of an ellipse cuts the curve in Q; the tangent at P is met by the other common tangent, which touches the curves at E and F, in T; if PQ meet TEF in O, $TEOF$ is cut harmonically.

26. If E is the centre of curvature at the point P of a parabola,

$$SE^2 + 3 \,.\, SP^2 = PE^2.$$

27. Find the locus of the foci of the parabolas which have a given circle as circle of curvature, at a given point of that circle.

28. Two parabolas, whose latera recta have a constant ratio, and whose foci are two given points A, B, have a contact of the second order at P. Shew that the locus of P is a circle.

29. If the fixed straight line PQ is the chord of an ellipse, and is also the diameter of curvature at P, prove that the locus of the centre of the ellipse is a rectangular hyperbola, the transverse axis of which is coincident in direction with PQ, and equal in length to one-half of PQ.

CHAPTER VIII.

ORTHOGONAL PROJECTIONS.

170. DEF. The projection of a point on a plane is the foot of the perpendicular let fall from the point on the plane.

If from all points of a given curve perpendiculars be let fall on a plane, the curve formed by the feet of the perpendiculars is the projection of the given curve.

The projection of a straight line is also a straight line, for it is the line of intersection with the given plane of a plane through the line perpendicular to the given plane.

Parallel straight lines project into parallel lines, for the projections are the lines of intersection of parallel planes with the given plane.

171. PROP. I. *Parallel straight lines, of finite lengths, are projected in the same ratio.*

That is, if ab, pq be the projections of the parallel lines AB, PQ,

$$ab : AB :: pq : PQ.$$

For, drawing AC parallel to ab and meeting Bb in C, and PR parallel to pq and meeting Qq in R, ABC and PQR are similar triangles; therefore

$$AC : AB :: PR : PQ,$$

and $$AC = ab,\ PR = pq.$$

172. PROP. II. *The projection of the tangent to a curve at any point is the tangent to the projection of the curve at the projection of the point.*

For if p, q be the projections of the two points P, Q of a curve, the line pq is the projection of the line PQ, and when the line PQ turns round P until Q coincides with P, pq turns round p until q coincides with p, and the ultimate position of pq is the tangent at p.

173. Prop. III. *The projection of a circle is an ellipse.* Let aba' be the projection of a circle ABA'.

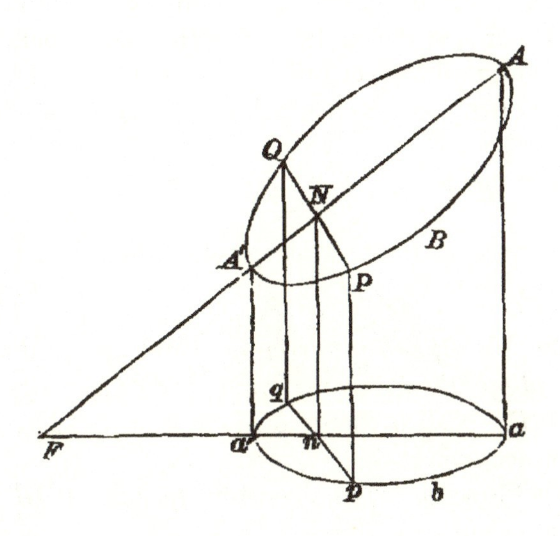

Take a chord PQ parallel to the plane of projection, then its projection $pq = PQ$.

Let the diameter ANA' perpendicular to PQ meet in F the plane of projection, and let $aa'F$ be the projection of $AA'F$.

Then aa' bisects pq at right angles in the point n, and

$$an : AN :: aF : AF,$$

$$a'n : A'N :: aF : AF;$$

$$\therefore AN \,.\, NA' : an \,.\, na' :: AF^2 : aF^2;$$

but
$$AN \,.\, NA' = PN^2 = pn^2,$$

$$\therefore pn2 : an \,.\, na' :: AF^2 : aF^2,$$

and the curve apa' is an ellipse, having its axes in the ratio of

$$aF : AF, \quad \text{or of} \quad aa' : AA'.$$

Moreover, since we can place the circle so as to make the ratio of aa' to AA' whatever we please, an ellipse of any eccentricity can be obtained.

In this demonstration we have assumed only the property of the principal diameters of an ellipse. Properties of other diameters can be obtained by help of the preceding theorems, as in the following instances.

174. PROP. IV. *The locus of the middle points of parallel chords of an ellipse is a straight line.*

For, projecting a circle, the parallel chords of the ellipse are the projections of parallel chords of the circle, and as the middle points of these latter lie in a diameter of the circle, the middle points of the chords of the ellipse lie in the projection of the diameter, which is a straight line, and is a diameter of the ellipse.

Moreover, the diameter of the circle is perpendicular to the chords it bisects; hence

Perpendicular diameters of a circle project into conjugate diameters of an ellipse.

175. PROP. V. *If two intersecting chords of an ellipse be parallel to fixed lines, the ratio of the rectangles contained by their segments is constant.*

Let OPQ, ORS be two chords of a circle, parallel to fixed lines, and opq, ors their projections.

Then $OP\,.\,OQ$ is to $op\,.\,oq$ in a constant ratio, and $OR\,.\,OS$ is to $or\,.\,os$ in a constant ratio; but

$$OP\,.\,OQ = OR\,.\,OS.$$

Therefore $op\,.\,oq$ is to $or\,.\,os$ in a constant ratio; and opq, ors are parallel to fixed lines.

176. PROP. VI. *If qvq' be a double ordinate of a diameter cp, and if the tangent at q meet cp produced in t,*

$$cv\,.\,ct = cp^2.$$

The lines qvq' and cp are the projections of a chord QVQ' of a circle which is bisected by a diameter CP, and t is the projection of T the point in which the tangent at Q meets CP produced.

But, in the circle,

$$CV\,.\,CT = CP^2,$$

or

$$CV : CP :: CP : CT;$$

and, these lines being projected in the same ratio, it follows that

$$cv : cp :: cp : ct,$$

or

$$cv\,.\,ct = cp^2.$$

Hence it follows that tangents to an ellipse at the ends of any chord meet in the diameter conjugate to the chord.

The preceding articles will shew the utility of the method in dealing with many of the properties of an ellipse.

The student will find it useful to prove, by orthogonal projections, the theorems of Arts. 58, 69, 74, 75, 78, 79, 80, 82, 83, 89, 90, and 92.

177. PROP. VII. *An ellipse can be projected into a circle.*

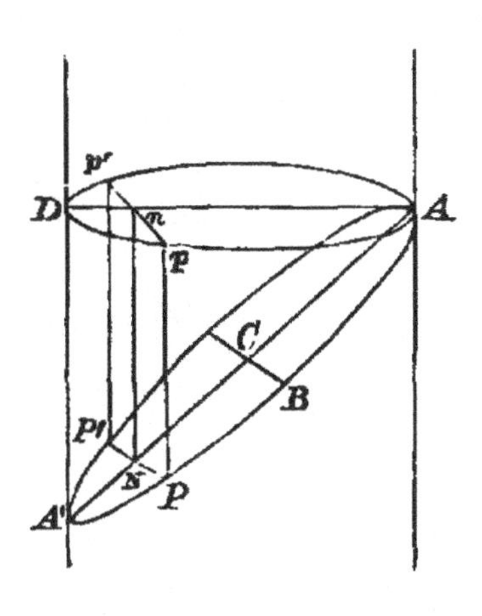

This is really the converse of Art. 173, but we give a construction for the purpose.

Draw a plane through AA', the transverse axis, perpendicular to the plane of the ellipse, and in this plane describe a circle on AA' as diameter. Also take the chord AD, equal to the conjugate axis, and join $A'D$, which is perpendicular to AD.

Through AD draw a plane perpendicular to $A'D$, and project a principal chord PNP' on this plane.

Then $$PN^2 : AN \,.\, NA' :: BC^2 : AC^2.$$

But $$PN = pn,$$

$$An : AN :: AD : AA'$$

$$:: BC : AC,$$

and $$Dn : A'N :: BC : AC.$$

Hence $$An \,.\, nD : AN.NA' :: BC^2 : AC^2,$$

and therefore $$pn^2 = An \,.\, nD,$$

and the projection ApD is a circle.

This theorem, in the same manner as that of Art. 173, may be employed in deducing properties of oblique diameters and oblique chords of an ellipse.

178. *If any figures in one plane be projected on another plane, the areas of the projections will all be in the same ratio to the areas of the figures themselves.*

Let BAD be the plane of the figures, and let them be projected on the plane CAD, C being the projection of the point B, and BAD being a right angle.

Taking a rectangle $EFGH$, the sides of which are parallel and perpendicular to AD, the projection is $efgh$, and it is clear that the ratio of the areas of these rectangles is that of AC to AB.

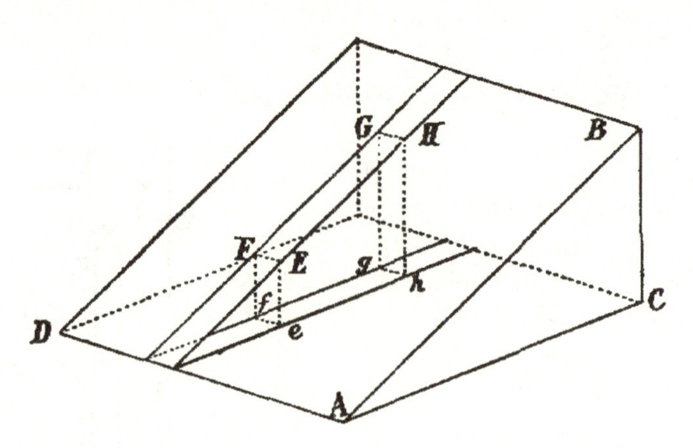

Now the area of any curvilinear figure in the plane BAD is the sum of the areas of parallelograms such as $EFGH$, which are inscribed in the figure, if we take the widths, such as EF, infinitesimally small.

It follows that the area of the projection of the figure is to the area of the figure itself in the ratio of AC to AB.

As an illustration, let a square be drawn circumscribing a circle, and project the figure on any plane. The square projects into tangents parallel to conjugate diameters of the ellipse which is the projection of the circle.

The area of the parallelogram thus formed is the same whatever be the position of the square, and we thus obtain the theorem of Art. 87.

179. It follows that maxima and minima areas project into maxima and minima areas. For example, the greatest triangle which can be inscribed in a circle is an equilateral triangle.

Projecting this figure we find that the triangle of maximum area inscribed in an ellipse is such that the tangent at each angular point is parallel to the opposite side, and that the centre of the ellipse is the point of intersection of the lines joining the vertices of the triangle with the middle points of the opposite sides.

180. PROP. VII. *The projection of a parabola is a parabola.*

For if PNP' be a principal chord, bisected by the axis AN, the projection pnp' will be bisected by the projection an.

Moreover $pn : PN$ will be a constant ratio, as also will be $an : AN$.

And $$PN^2 = 4AS \,.\, AN.$$

Hence pn^2 will be to $4AS \,.\, an$ in a constant ratio, and the projection is a parabola, the tangent at a being parallel to pn.

181. Prop. VIII. *An hyperbola can be always projected into a rectangular hyperbola.*

For the asymptotes can be projected into two straight lines cl, cl' at right angles, and if PM, PN be parallels to the asymptotes from a point P of the curve, $PM \,.\, PN$ is constant.

But $pm : PM$ and $pn : PN$ are constant ratios;

$$\therefore pm \,.\, pn \text{ is constant.}$$

And since pm and pn are perpendicular respectively to cl and cl', it follows that the projection is a rectangular hyperbola.

The same proof evidently shews that any projection of an hyperbola is also an hyperbola.

EXAMPLES.

1. A parallelogram is inscribed in a given ellipse; shew that its sides are parallel to conjugate diameters, and find its greatest area.

2. TP, TQ are tangents to an ellipse, and CP', CQ' are parallel semi-diameters; PQ is parallel to $P'Q'$.

3. If a straight line meet two concentric similar and similarly situated ellipses, the portions intercepted between the curves are equal.

4. Find the locus of the point of intersection of the tangents at the extremities of pairs of conjugate diameters of an ellipse.

5. Find the locus of the middle points of the lines joining the extremities of conjugate diameters.

6. If a tangent be drawn at the extremity of the major axis meeting two equal conjugate diameters CP, CD produced in T and t; then $PD^2 = 2AT^2$.

7. If a chord AQ drawn from the vertex be produced to meet the minor axis in O, and CP be a semi-diameter parallel to it, then $AQ \,.\, AO = 2CP^2$.

8. OQ, OQ' are tangents to an ellipse from an external point O, and OR is a diagonal of the parallelogram of which OQ, OQ' are adjacent sides; prove that if R be on the ellipse, O will lie on a similar and similarly situated concentric ellipse.

9. AB is a given chord of an ellipse, and C any point in the ellipse; shew that the locus of the point of intersection of lines drawn from A, B, C to the middle points of the opposite sides of the triangle ABC is a similar ellipse.

10. CP, CD are conjugate semi-diameters of an ellipse; if an ellipse, similar and similarly situated to the given ellipse, be described on PD as diameter, it will pass through the centre of the given ellipse.

11. Parallelograms are inscribed in an ellipse and one pair of opposite sides constantly touch a similar, similarly situated and concentric ellipse; shew that the remaining pair of sides are tangents to a third ellipse and the square on a principal semi-axis of the original ellipse is equal to the sum of the squares on the corresponding semi-axes of the other two ellipses.

12. Find the locus of the middle point of a chord of an ellipse which cuts off a constant area from the curve.

13. Find the locus of the middle point of a chord of a parabola which cuts off a constant area from the curve.

14. A parallelogram circumscribes an ellipse, touching the curve at the extremities of conjugate diameters, and another parallelogram is formed by joining the points where its diagonals meet the ellipse: prove that the area of the inner parallelogram is half that of the outer one.

If four similar and similarly situated ellipses be inscribed in the spaces between the outer parallelogram and the curve, prove that their centres lie in a similar and similarly situated ellipse.

15. About a given triangle PQR is circumscribed an ellipse, having for centre the point of intersection (C) of the lines from P, Q, R bisecting the opposite sides, and PC, QC, RC are produced to meet the curve in P', Q', R'; shew that, if tangents be drawn at these points, the triangle so formed will be similar to PQR, and four times as great.

16. The locus of the middle points of all chords of an ellipse which pass through a fixed point in an ellipse similar and similarly situated to the given ellipse, and with its centre in the middle point of the line joining the given point and the centre of the given ellipse.

17. PT, pt are tangents at the extremities of any diameter Pp of an ellipse; any other diameter meets PT in T and its conjugate meets pt in t; also any tangent meets PT in T' and pt in t'; shew that $PT : PT' :: pt' : pt$.

18. From the ends P, D of conjugate diameters of an ellipse lines are drawn parallel to any tangent line; from the centre C any line is drawn cutting these lines and the tangent in p, d, t, respectively; prove that $Cp^2 + Cd^2 = Ct^2$.

19. If CP, CD be conjugate diameters of an ellipse, and if BP, BD be joined, and also AD, $A'P$, these latter intersecting in O, the figure $BDOP$ will be a parallelogram.

20. T is a point on the tangent at a point P of an ellipse, so that a perpendicular from T on the focal distance SP is of constant length; shew that the locus of T is a similar, similarly situated and concentric ellipse.

21. Q is a point in one asymptote, and q in the other. If Qq move parallel to itself, find the locus of intersection of tangents to the hyperbola from Q and q.

22. Tangents are drawn to an ellipse from an external point T. The chord of contact and the major axis, or these produced, intersect in K, and TN is drawn perpendicular to the major axis. Prove that

$$CN \,.\, CK = CA^2.$$

23. Q is a variable point on the tangent at a fixed point P of an ellipse and R is taken so that $PQ = QR$. If the other tangent from Q meet the ellipse in K, prove that RK passes through a fixed point.

24. If through any point on an ellipse there be drawn lines conjugate to the sides of an inscribed triangle they will meet the sides in three points in a straight line.

25. PCP' is a diameter of an ellipse, and a chord PQ meets the tangent at P' in R. Prove that PQ, PR have the parallel diameter for a mean proportional.

26. If AOA', BOB' are conjugate diameters of an ellipse, and if AP and BQ are parallel chords, $A'Q$ and $B'P$ are parallel to conjugate diameters.

27. If the tangents at the ends of a chord of an hyperbola meet in T, and TM, TM' be drawn parallel to the asymptotes to meet them in M, M', then MM' is parallel to the chord.

28. If a windmill in a level field is working uniformly on a sunny day, the speed of the end of the shadow of one sail varies as the length of the shadow of the next sail.

29. Spheres are drawn passing through a fixed point and touching two fixed planes. Prove that the points of contact lie on two circles, and that the locus of the centre of the sphere is an ellipse.

If the angle between the planes is the angle of an equilateral triangle, prove that the distance between the foci of the ellipse is half the major axis.

CHAPTER IX.

OF CONICS IN GENERAL.

The Construction of a Conic.

182. The method of construction, given in Chapter I., can be extended in the following manner.

Let fSn be any straight line drawn through the focus S, and draw Ax from the vertex parallel to fS, and meeting the directrix in x.

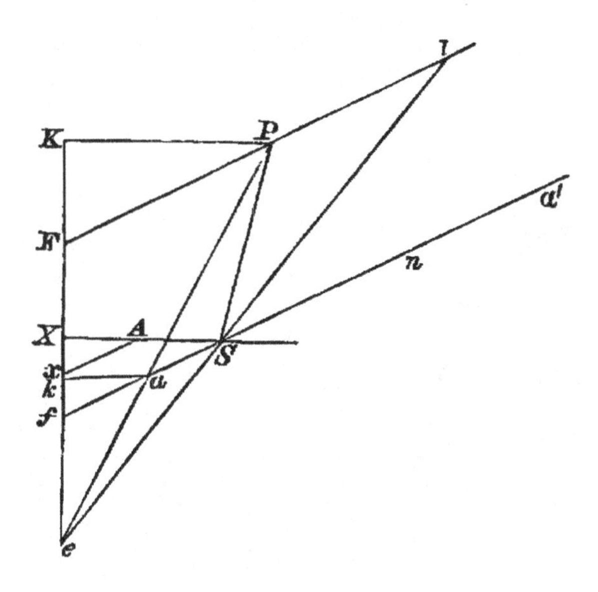

Divide the line fSn in a and a' so that

$$Sa : af :: Sa' : a'f :: SA : Ax;$$

then a and a' are points on the curve, for, if ak be the perpendicular on the directrix,

$$ak : af :: AX : Ax,$$

and therefore $$Sa : ak :: SA : AX.$$

Take any point e in the directrix, draw the lines eSl, ea through S and a, and draw SP making the angle PSl equal to lSn.

Through P draw FPl parallel to fS, and meeting eS produced in l,

then $$Pl = SP,$$

and $$Pl : PF :: Sa : af;$$

$$\therefore SP : PF :: Sa : af,$$

and $$SP : PK :: Sa : ak;$$

therefore P is a point in the curve.

183. The construction for the point a gives a simple proof that the tangent at the vertex is perpendicular to the axis. For when the angle ASa is diminished, Sa approaches to equality with SA, and therefore the angle aAS is ultimately a right angle.

184. Prop. I. *To find the points in which a given straight line is intersected by a conic of which the focus, the directrix, and the eccentricity are given.*

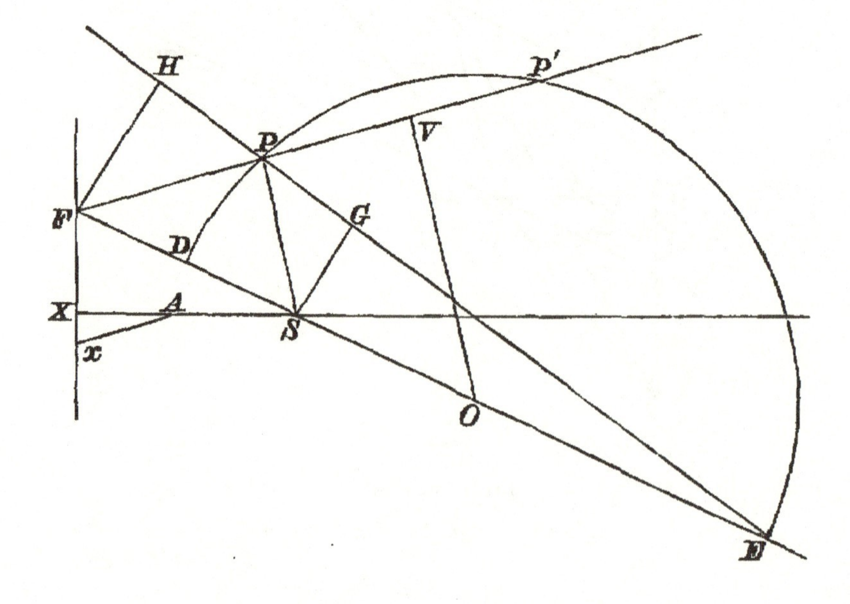

Let FPP' be the straight line, and draw Ax parallel to it. Join FS, and find the points D and E such that

$$SD : DF :: SE : EF :: SA : Ax.$$

Describe the circle on DE as diameter, and let it intersect the given line in P and P'.

Join DP, EP and draw SG, FH at right angles to EP.

Then DPE, being the angle in a semicircle, is a right angle, and DP is parallel to SG and FH.

Hence
$$\begin{aligned} SG : FH &:: SE : EF \\ &:: SD : DF \\ &:: PG : PH; \end{aligned}$$

therefore the angles SPG, FPH are equal, and therefore PD bisects the angle SPF.

Hence $SP : PF :: SD : DF :: SA : Ax$,
and P is a point in the curve.

Similarly P' is also a point in the curve, and the perpendicular from O, the centre of the circle, on FPP' meets it in V, the middle point of the chord PP'.

Since
$$SE : EF :: SA : Ax$$
and
$$SD : DF :: SA : Ax;$$
$$\therefore SE - SD : DE :: SA : Ax,$$
or
$$SO : OD :: SA : Ax,$$
a relation analogous to
$$SC : AC :: SA : AX.$$

We have already shewn, for each conic, that the middle points of parallel chords lie in a straight line; the following article contains a proof of the theorem which includes all the three cases.

185. Prop. II. *To find the locus of the middle points of a system of parallel chords.*

Let $P'P$ one of the chords be produced to meet the directrix in F, draw Ax parallel to FP, and divide FS so that

$$SD : DF :: SE : EF :: SA : Ax;$$

then, as in the preceding article, the perpendicular OV upon PP' from O, the middle point of DE, bisects PP'.

Draw the parallel focal chord aSa'; then Oc parallel to the directrix bisects aa' in c. Also draw SG perpendicular to the chords, and meeting the directrix in G.

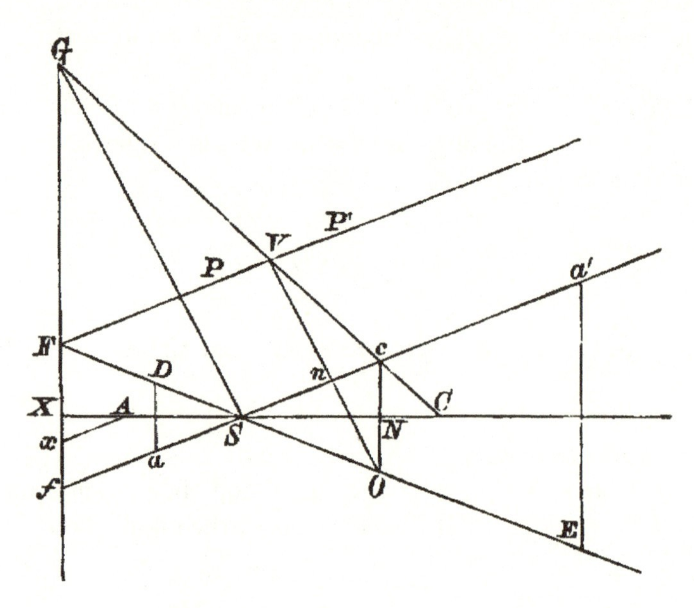

Then, if OV meet aa' in n,

$$Vn : nO :: SF : SO,$$
$$:: Sf : Sc,$$

and, since ncO, SGf are similar triangles,

$$nO : nc :: SG : Sf;$$
$$\therefore Vn : nc :: SG : Sc,$$

and the line Vc passes through G.

The straight line Gc is therefore the locus of the middle points of all chords parallel to aSa'.

The ends of the diameter GC may be found by the construction of the preceding article.

186. When the conic is a parabola, $SA = AX$,

and $$Sa : af :: AX : Ax$$
$$:: SX : Sf.$$

So $$Sa' : a'f :: SX : Sf;$$

$$\therefore Sc : ac :: SX : Sf,$$

and $$ac : cf :: SX : Sf.$$

Hence $$Sc : cf :: SX^2 : Sf^2$$
$$:: GX \,.\, Xf : Gf \,.\, fX$$
$$:: GX : Gf;$$

and therefore Gc is parallel to SX, that is, the middle points of parallel chords of a parabola lie in a straight line parallel to the axis.

187. PROP. III. *To find the locus of the middle points of all focal chords of a conic.*

Taking the case of a central conic, and referring to the figure of the preceding article, let Oc meet SC in N;

then $$cN : NS :: fX : SX,$$

and $$cN : NC :: GX : CX;$$

$$\therefore cN^2 : SN \,.\, NC :: fX \,.\, GX : SX \,.\, CX$$
$$:: SX^2 : SX \,.\, CX.$$

Hence it follows that the locus of c is an ellipse of which SC is the transverse axis, and such that the squares of its axes are as $SX : CX$, or (Cor. Art. 63) as $BC^2 : AC^2$.

Hence the locus of c is similar to the conic itself.

EXAMPLES.

1. If an ordinate, PNP', to the transverse axis meet the tangent at the end of the latus rectum in T,

$$SP = TN, \text{ and } TP \,.\, TP' = SN^2.$$

2. A focal chord PSQ of a conic section is produced to meet the directrix in K, and KM, KN are drawn through the feet of the ordinates PM, QN of P and Q. If KN produced meet PN produced in R, prove that

$$PR = PM.$$

3. The tangents at P and Q, two points in a conic, intersect in T; if through P, Q, chords be drawn parallel to the tangents at Q and P, and intersecting the conic in p and q respectively, and if tangents at p and q meet in T, shew that Tt is a diameter.

4. Two tangents TP, TP' are drawn to a conic intersecting the directrix in F, F'.

If the chord PQ cut the directrix in R, prove that

$$SF : SF' :: RF : RF'.$$

5. The chord of a conic PP' meets the directrix in K, and the tangents at P and P' meet in T; if RKR', parallel to ST, meet the tangents in R and R',

$$KR = KR'.$$

6. The tangents at P and P', intersecting in T, meet the latus rectum in D and D'; prove that the lines through D and D', respectively perpendicular to SP and SP', intersect in ST.

7. If P, Q be two points on a conic, and p, q two points on the directrix such that pq subtends at the focus half the angle subtended by PQ, either Pp and Qq or Pq and Qp meet on the curve.

8. A chord PP' of a conic meets the directrix in F, and from any point T in PP', TLL' is drawn parallel to SF and meeting SP, SP' in L and L'; prove that the ratio of SL or SL' to the distance of T from the directrix is equal to the ratio of $SA : AX$.

9. If an ellipse and an hyperbola have their axes coincident and proportional, points on them equidistant from one axis have the sum of the squares on their distances from the other axis constant.

10. If Q be any point in the normal PG, QR the perpendicular on SP, and QM the perpendicular on PN,

$$QR : PM :: SA : AX.$$

11. Given a focus of a conic section inscribed in a triangle, find the points where it touches the sides.

12. PSQ is any focal chord of a conic section; the normals at P and Q intersect in K, and KN is drawn perpendicular to PQ; prove that PN is equal to SQ, and hence deduce the locus of N.

13. Through the extremity P, of the diameter PQ of an ellipse, the tangent TPT' is drawn meeting two conjugate diameters in T, T'. From P, Q the lines PR, QR are drawn parallel to the same conjugate diameters. Prove that the rectangle under the semi-axes of the ellipse is a mean proportional between the triangles PQR and CTT'.

14. Shew that a conic may be drawn touching the sides of a triangle, having one focus at the centre of the circumscribing circle, and the other at the orthocentre.

15. The perpendicular from the focus of a conic on any tangent, and the central radius to the point of contact, intersect on the directrix.

16. AB, AC are tangents to a conic at B, and C, and $DEGF$ is drawn from a point D in AC, parallel to AB and cutting the curve in E and F, and BC in G; shew that

$$DG^2 = DE \,.\, DF.$$

17. A diameter of a parabola, vertex F, meets two tangents in D and E and their chord of contact is G, shew that

$$FG^2 = ED \,.\, FE.$$

18. P and Q are two fixed points in a parabola, and from any other point R in the curve, RP, RQ are drawn cutting a fixed diameter, vertex E, in B and C; prove that the ratio of EB to EC is constant.

19. If the normal at P meet the conjugate axis in g, and gk be perpendicular to SP, Pk is constant; and if kl, parallel to the transverse axis, meet the normal at P in l, kl is constant.

20. A system of conics is drawn having a common focus S and a common latus rectum LSL'. A fixed straight line through S intersects the conics, and at the points of intersection normals are drawn. Prove that the envelope of each of these normals is a parabola whose focus lies on LSL', and which has the given line as tangent at the vertex.

CHAPTER X.

ELLIPSES AS ROULETTES AND GLISSETTES.

188. *If a circle rolls on the inside of the circumference of a circle of double its radius, any point in the area of the rolling circle traces out an ellipse.*

Let C be the centre of the rolling circle, E the point of contact.

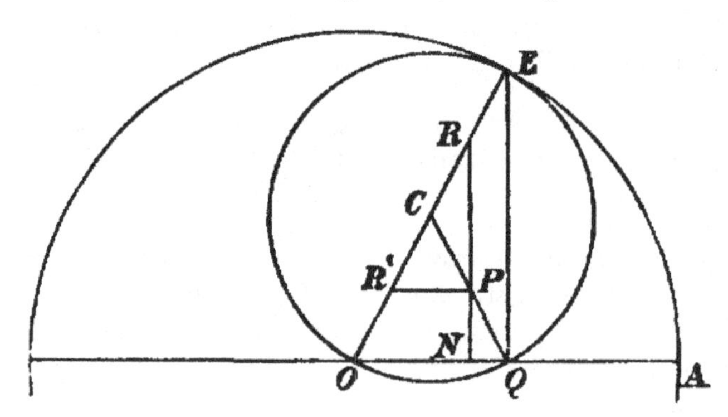

Then, if the circle meet in Q a fixed radius OA of the fixed circle, the angle ECQ is twice the angle EOA, and therefore the arcs EQ, EA are equal.

Hence, when the circles touch at A, the point Q of the rolling circle coincides with A, and the subsequent path of Q is the diameter through A.

Let P be a given point in the given radius CQ, and draw RPN perpendicular to OA, and PR' parallel to OA.

Then, OQE being a right angle, EQ is parallel to RP and therefore $CR = CP = CR'$, so that OR and OR' are constant.

Also $$PN : RN :: PQ : OR;$$

therefore, the locus of R being a circle, the locus of P is an ellipse, whose axes are as $PQ : OR$.

But OR is clearly the length of one semi-axis, and PQ or OR' is therefore the length of the other, OR, OR' being equal to $OC + CP$ and $OC - CP$.

189. Properties of the ellipse are deducible from this construction.

Thus, as the circle rolls, the point E is instantaneously at rest, and the motion of P is therefore at right angles to EP, *i.e.* producing EP to F, in the direction FO.

Therefore, drawing PT parallel to OF, PT is the tangent, and PF the normal.

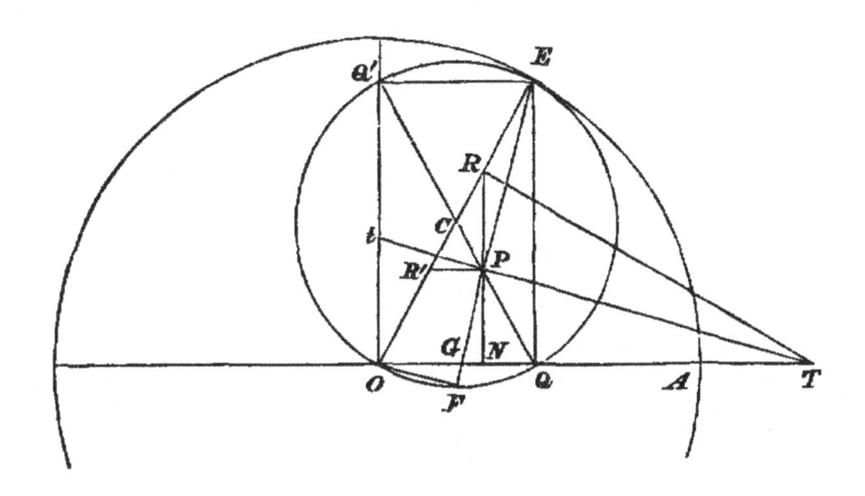

The angles EPT, EQT being right angles, the points E, P, Q, T are concyclic; but the circle through QPE clearly passes through R; therefore the angle ERT and consequently the angle ORT is a right angle,

and $$ON : OR :: OR : OT,$$

or $$ON \,.\, OT = OR^2,$$

which is the theorem of Art. 74.

Again, since $EQ't$ and EPt are right angles, E, Q', t, P are concyclic; but the circle through $EQ'P$ clearly passes through R'; therefore the angle $ER't$ and consequently the angle $OR't$ is a right angle, and

$$PN : OR' :: OR' : Ot,$$

or $$PN \,.\, Ot = OR'^2,$$

which is the theorem of Art. 75.

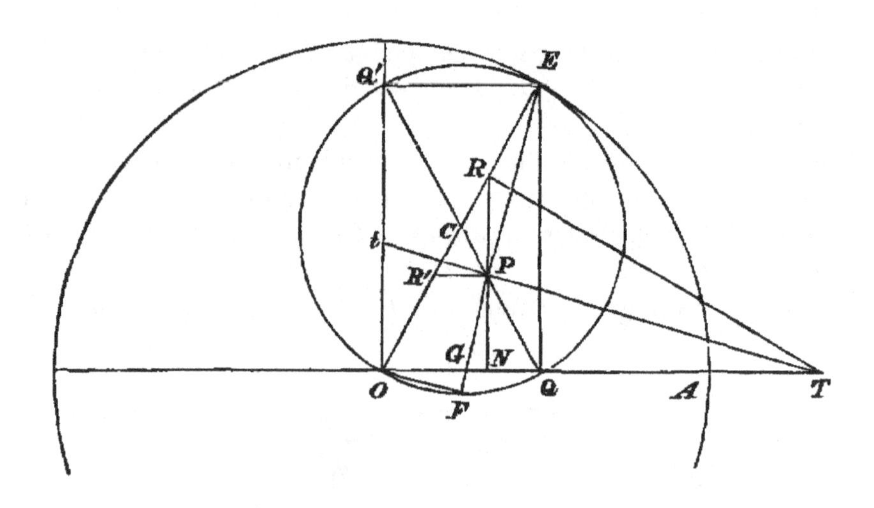

Further, if PF meet OQ in G, the angles PQG, PFQ are equal, being on equal bases EQ, OQ';

$$\therefore PG : PQ :: PQ : PF,$$

or

$$PG \,.\, PF = PQ^2 = OR'^2,$$

which is the first of theorems of Art. 77.

And again, if PGF produced meet $Q'O$ produced in g, the angles $PQ'g$, PFQ' are equal, being on equal bases QO, EQ'; and the angle $Q'Pg$ is common to the two triangles $PQ'g$, PFQ'.

Therefore these triangles are similar, and

$$Pg : PQ' :: PQ' : PF,$$

or

$$Pg \,.\, PF = PQ'^2.$$

But

$$PQ' = ER' = OR;$$

$$\therefore Pg \,.\, PF = OR^2,$$

which is the second theorem of Art. 77.

190. If the carried point P is outside the circle the line PNR, perpendicular to OA, will meet OE produced in R, and CR will be equal to CP, so that OR will be constant and the locus of R will be a circle.

Also, the triangles PQN, RON being similar, we shall have

$$PN : RN :: PQ : OR,$$

so that the locus of P will be an ellipse, the semi-axes of which will be $CP + OC$ and $CP - OC$.

191. The fact that a point on the circumference of the rolling circle oscillates in a straight line is utilized in the construction of Wheatstone's Photometer.

By help of machinery a metallic circle, about an inch in diameter, is made to roll rapidly round the inside of a circle of double this diameter, and carries a small bright bead which is fastened to its circumference.

If this machine is held between two candles or other sources of light, so that the line of oscillation of the bead is equidistant from the candles, two bright lines will be seen in close contiguity, and it is easy to form an estimate of their comparative brightnesses.

If bright beads are fastened to points in the area of the rolling circle not on the circumference, and the machine be held near sources of light, the appearance, when the circle is made to rotate rapidly, will be that of a number of bright concentric ellipses.

192. *A given straight line has its ends moveable on two straight lines at right angles to each other; the path of any given point in the moving line is an ellipse.*

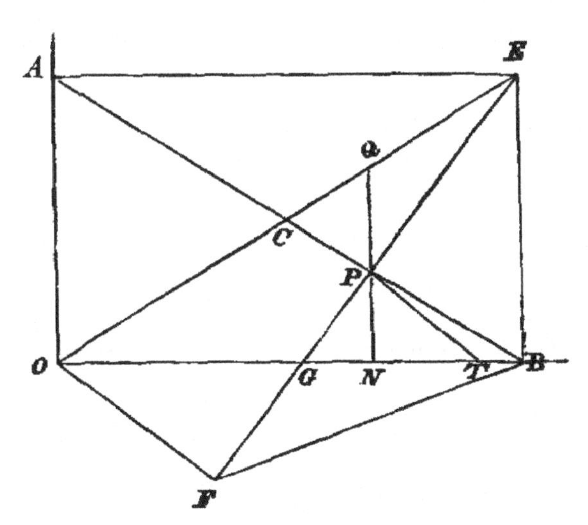

Let P be the point in the moving line AB, and C the middle point of AB.

Let the ordinate NP, produced if necessary, meet OC in Q; then $CQ = CP$ and $OQ = AP$, so that the locus of Q is a circle.

Also
$$PN : QN :: PB : OQ$$
$$:: PB : PA;$$

therefore the locus of P is an ellipse, and its semi-axes are equal to AP and BP.

193. The theorem of Art. 188 is at once reducible to this case, for, taking the figure of Art. 189, QPQ' is a diameter of the rolling circle and is therefore of constant length, and the points Q and Q' move along fixed straight lines at right angles to each other; the locus of P is therefore an ellipse of which $Q'P$ and PQ are the semi-axes.

194. From this construction also properties of the tangent and normal are deducible.

Complete the rectangle $OAEB$; then, since the directions of motion of A and B are respectively perpendicular to EA and EB, the state of motion of the line AB may be represented by supposing that the triangle EAB is turning round the point E.

Hence it follows that EP is the normal to the locus of P, and that PT perpendicular to EP is the tangent.

Let OF, parallel to PT, meet EP in F; then O, F, B, E are concyclic;

$$\therefore \text{ the angle } PFB = EOB = PBG,$$

and the triangles PGB, PFB are similar.

Hence
$$PG : PB :: PB : PF,$$
or
$$PG \,.\, PF = PB^2,$$
where PB is equal to the semi-conjugate axis.

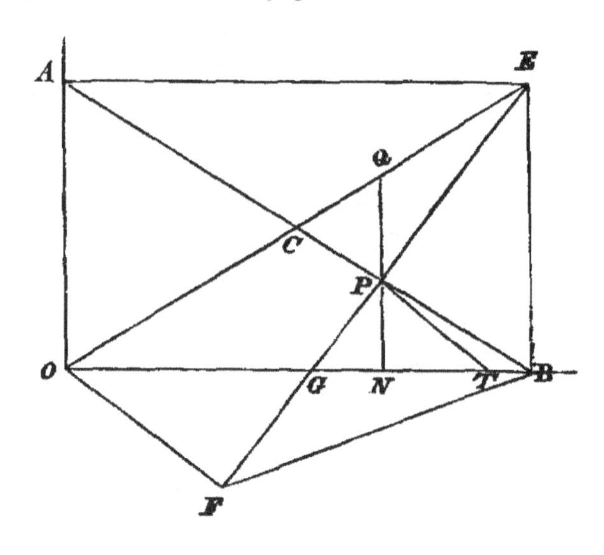

Similarly, by joining AF, it can be shewn that

$$Pg \,.\, PF = PA^2,$$

g being the point of intersection of PG and AO.

Again, since EPT, EBT are right angles, B, T, P, E are concyclic, and Q is clearly concyclic with B, P, E; so that TQE is a right angle.

Hence OQN and OQT are similar triangles, and

$$ON : OQ :: OQ : OT,$$

or

$$ON \,.\, OT = PA^2,$$

where PA is equal to the semi-transverse axis.

195. Observing that F, O, A, E, B are concyclic, we have

$$PF \,.\, PE = PA \,.\, PB;$$

$\therefore PE$ is equal to the semi-diameter conjugate to OP.

This suggests a construction for the solution of the problem,

Having given a pair of conjugate diameters of an ellipse, it is required to determine the position and magnitudes of the principal axes.

Taking OP and OD as the given semi-conjugate diameters, draw PF perpendicular to OD, and, in FP produced, take PE equal to OD.

Join OE, bisect it in C, and in CE take CQ equal to CP.

Then OB, OA, drawn perpendicular and parallel to PQ, and meeting CP in B and A, will be the directions of the axes, and their lengths will be AP and PB.

196. *If a given triangle AQB move in its own plane so that the extremities A, B, of its base AB move on two fixed straight lines at right angles to each other, the path of the point Q is an ellipse.*

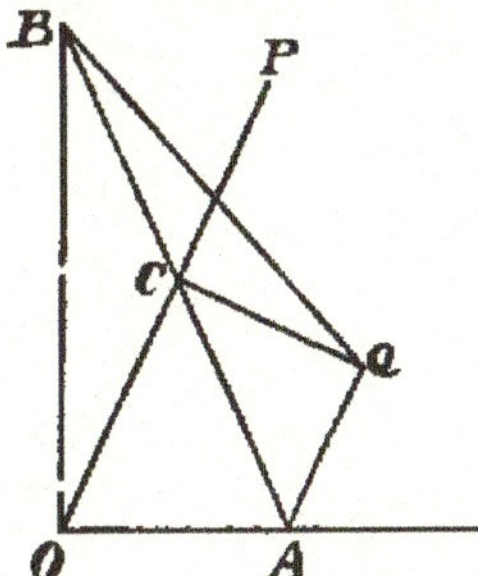

If O be the point of intersection of the fixed lines, and C the middle point of AB, the angles COB, CBO are equal, so that, as AB slides, the line CB, and therefore also the line CQ, turns round as fast as CO, but in the contrary direction.

Produce OC to P, making $CP = CQ$; then the locus of P is a circle the radius of which is equal to $OC + CQ$.

There is clearly one position of AB for which the points O, C, and Q are in one straight line.

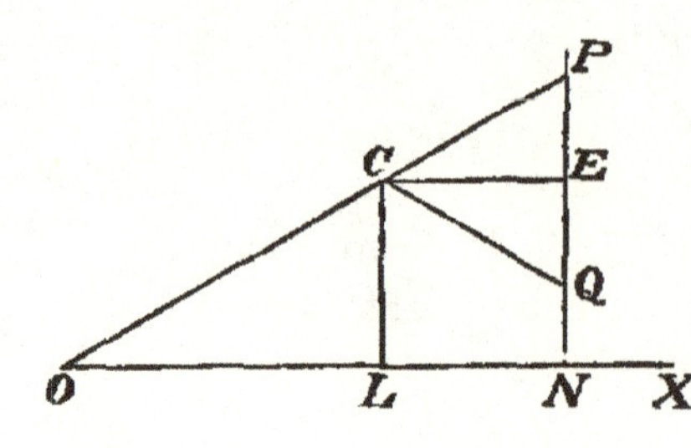

Let OX be this straight line, and let OC, CQ, be any other corresponding positions of the lines;
then, if CE is parallel to OX, CE bisects the angle PCQ, and, drawing PQN and CL perpendicular to OX,

$$QN = CL - PE, \quad PN = CL + PE,$$

hence
$$QN : PN :: OC - CP : OC + CP$$
$$:: OC - CQ : OC + CQ,$$

and $\therefore$ the locus of Q is an ellipse of which the semi-axes are $OC + CQ$ and $OC - CQ$.

If the straight lines through A and B perpendicular to OA and OB meet in K, the point K is the instantaneous centre of rotation. The normal to the path of Q is therefore QK and the tangent is the straight line through Q perpendicular to QK.

197. *Elliptic Compasses.* If two fine grooves, at right angles to each other, be made on the plane surface of a plate of wood or metal, and if two pegs, fastened to a straight rod, be made to move in these grooves, then a pencil attached to any point of the rod will trace out an ellipse.

By fixing the pencil at different points of the rod, we can obtain ellipses of any eccentricity, but of dimensions limited by the lengths of the rod and the grooves.

Burstow's Elliptograph.

OE is a groove in a stand which can be fixed to the paper or drawing board, and OA, OB are rods jointed at A, so that the end B can slip along the groove, while AO turns round the fixed end O.

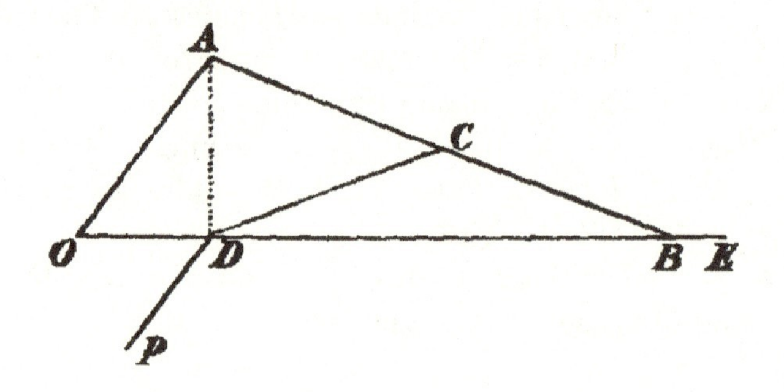

C is the middle point of AB, CD is a rod, the length of which is half that of AB, and the end D can slide along the groove.

It follows that the angle ADB is always a right angle.

A rod DP is taken of any convenient length, and, by means of a chain round the triangle ADC, is made to move so as to be always parallel to OA.

If the end B be moved along the groove, the end P will trace out an ellipse of which O is the centre, and the lengths of its semi-axes will be the length of DP and of the difference between the lengths of OA and DP. This can be seen by drawing a line OF perpendicular to OE, and producing DP to meet it in F. The motion will be that of a rod of length OA sliding between OE and OF. See Dyck, *Katalog der mathematischen Instrumente*, München, 1892.

MISCELLANEOUS PROBLEMS. I.

1. On a plane field the crack of the rifle and the thud of the ball striking the target are heard at the same instant; find the locus of the hearer.

2. PQ, $P'Q'$ are two focal chords of a parabola, and PR, parallel to $P'Q'$, meets in R the diameter through Q; prove that

$$PQ \,.\, P'Q' = PR^2.$$

3. CP and CD are conjugate semi-diameters of an ellipse; PQ is a chord parallel to one of the axes; shew that DQ is parallel to one of the straight lines which join the ends of the axes.

4. A line cuts two concentric, similar and similarly situated ellipses in P, Q, q, p. If the line move parallel to itself, $PQ \,.\, Qp$ is constant.

5. The portion of a tangent to an hyperbola intersected between the asymptotes subtends a constant angle at the focus.

6. If a circle be described passing through any point P of a given hyperbola and the extremities of the transverse axis, and the ordinate NP be produced to meet the circle in Q, the locus of Q is an hyperbola.

7. PQ is one of a series of chords inclined at a constant angle to the diameter AB of a circle; find the locus of the intersection of AP, BQ.

8. If from a point T in the director circle of an ellipse tangents TP, TP' be drawn, the line joining T with the intersection of the normals at P and P' passes through the centre.

9. The points, in which the tangents at the extremities of the transverse axis of an ellipse are cut by the tangent at any point of the curve, are joined, one with each focus; prove that the point of intersection of the joining lines lies in the normal at the point.

10. Having given a focus, the eccentricity, a point of the curve, and the tangent at the point, shew that in general two conics can be described.

11. A parabola is described with its focus at one focus of a given central conic, and touches the conic; prove that its directrix will touch a fixed circle.

12. The extremities of the latera recta of all conics which have a common transverse axis lie on two parabolas.

13. The tangent at a moveable point P of a conic intersects a fixed tangent in Q, and from S a straight line is drawn perpendicular to SQ and meeting in R the tangent at P; prove that the locus of R is a straight line.

14. On all parallel chords of a circle a series of isosceles triangles are described, having the same vertical angle, and having their planes perpendicular to the plane of the circle. Find the locus of their vertices; and find what the vertical angle must be in order that the locus may be a circle.

15. A series of similar ellipses whose major axes are in the same straight line pass through two given points. Prove that the major axes subtend right angles at four fixed points.

16. From the centre of two concentric circles a straight line is drawn to cut them in P and Q; through P and Q straight lines are drawn parallel to two given lines at right angles to each other. Shew that the locus of their point of intersection is an ellipse.

17. A circle always passes through a fixed point, and cuts a given straight line at a constant angle, prove that the locus of its centre is an hyperbola.

18. The area of the triangle formed by three tangents to a parabola is equal to one half that of the triangle formed by joining the points of contact.

19. If a parabola be described with any point on an hyperbola for focus and passing through the foci of the hyperbola, shew that its axis will be parallel to one of the asymptotes.

20. S and H being the foci, P a point in the ellipse, if HP be bisected in L, and AL be drawn from the vertex cutting SP in Q, the locus of Q is an ellipse whose focus is S.

21. If the diagonals of a quadrilateral circumscribing an ellipse meet in the centre the quadrilateral is a parallelogram.

22. A series of ellipses pass through the same point, and have a common focus, and their major axes of the same length; prove that the locus of their centres is a circle. What are the limits of the eccentricities of the ellipses, and what does the ellipse become at the higher limit?

23. If S, H be the foci of an hyperbola, LL' any tangent intercepted between the asymptotes, $SL \,.\, HL = CL \,.\, LL'$.

24. Tangents are drawn to an ellipse from a point on a similar and similarly situated concentric ellipse; shew that if P, Q be the points of contact, A, A' the ends of the axis of the first ellipse, the loci of the intersections of AP, $A'Q$, and of AQ, $A'P$ are two ellipses similar to the given ellipses.

25. Draw a parabola which shall touch four given straight lines. Under what condition is it possible to describe a parabola touching five given straight lines?

26. A fixed hyperbola is touched by a concentric ellipse. If the curvatures at the point of contact are equal the area of the ellipse is constant.

27. A circle passes through a fixed point, and cuts off equal chords AB, CD from two given parallel straight lines; prove that the envelope of each of the chords AD, BC is a central conic having the fixed point for one focus.

28. A straight line is drawn through the focus parallel to one asymptote and meeting the other; prove that the part intercepted between the curve and the asymptote is one-fourth the transverse axis, and the part between the curve and the focus one-fourth the latus rectum.

29. PQ is any chord of a parabola, cutting the axis in L; R, R' are the two points in the parabola at which this chord subtends a right angle: if RR' be joined, meeting the axis in L', LL' will be equal to the latus rectum.

30. If two equal parabolas have the same focus, tangents at points angularly equidistant from the vertices meet on the common tangent.

31. A parabola has its focus at S, and PSQ is any focal chord, while PP', QQ' are two chords drawn at right angles to PSQ at its extremities; shew that the focal chord drawn parallel to PP' is a mean proportional between PP' and QQ'.

32. With the orthocentre of a triangle as centre are described two ellipses, one circumscribing the triangle and the other touching its sides; prove that these ellipses are similar, and their homologous axes at right angles.

33. $ABCD$ is a quadrilateral, the angles at A and C being equal; a conic is described about $ABCD$ so as to touch the circumscribing circle of ABC at the point B; shew that BD is a diameter of the conic.

34. The volume of a cone cut off by a plane bears a constant ratio to the cube, the edge of which is equal to the minor axis of the section.

35. A tangent to an ellipse at P meets the minor axis in t, and tQ is perpendicular to SP; prove that SQ is of constant length, and that if PM be the perpendicular on the minor axis, QM will meet the major axis in a fixed point.

36. Describe an ellipse with a given focus touching three given straight lines, no two of which are parallel and on the same side of the focus.

37. Prove that the conic which touches the sides of a triangle, and has its centre at the centre of the nine-point circle, has one focus at the orthocentre, and the other at the centre of the circumscribing circle.

38. From Q, the middle point of a chord PP' of an ellipse whose focus is S, QG is drawn perpendicular to PP' to meet the major axis in G; prove that

$$2 \,.\, SG : SP + SP' :: SA : AX.$$

39. A straight rod moves in any manner in a plane; prove that, at any instant, the directions of motion of all its particles are tangents to a parabola.

40. If from a point T on the auxiliary circle, two tangents be drawn to an ellipse touching it in P and Q, and when produced meeting the circle again in p, q; shew that the angles PSp and QSq are together equal to the supplement of PTQ.

41. Tangents at the extremities of a pair of conjugate diameters of an ellipse meet in T; prove that ST, $S'T$ meet the conjugate diameters in four concyclic points.

42. From the point of intersection of an asymptote and a directrix of an hyperbola a tangent is drawn to the curve; prove that the line joining the point of contact with the focus is parallel to the asymptote.

43. If a string longer than the circumference of an ellipse be always drawn tight by a pencil, the straight portions being tangents to the ellipse, the pencil will trace out a confocal ellipse.

44. D is any point in a rectangular hyperbola from which chords are drawn at right angles to each other to meet the curve. If P, Q be the middle points of these chords, prove that P, Q, D and the centre of the hyperbola are concyclic.

45. From a point T in the auxiliary circle tangents are drawn to an ellipse, touching it in P and Q, and meeting the auxiliary circle again in p and q; shew that the angle pCq is equal to the sum of the angles PSQ and $PS'Q$.

46. The angle between the focal distance and tangent at any point of an ellipse is half the angle subtended at the focus by the diameter through the point.

47. H is a fixed point on the bisector of the exterior angle A of the triangle ABC; a circle is described upon HA as chord cutting the lines AB, AC in P and Q; prove that PQ envelopes a parabola which has H for focus, and for tangent at the vertex the straight line joining the feet of the perpendiculars from H on AB and AC.

48. Tangents to an ellipse, foci S and H, at the ends of a focal chord PHP' meet the further directrix in Q, Q'. The parabola, whose focus is S, and directrix PP', touches PQ, $P'Q'$, in Q, Q'; it also touches the normals at P, P', and the minor axis, and has for the tangent at its vertex the diameter parallel to PP'.

49. S is a fixed point, and E a point moving on the arc of a given circle; prove that the envelope of the straight line through E at right angles to SE is a conic.

50. A circle passing through a fixed point S cuts a fixed circle in P, and has its centre at O; the lines which bisect the angle SOP all touch a conic of which S is a focus.

51. The tangent to an ellipse at P meets the directrix, corresponding to S, in Z: through Z a straight line ZQR is drawn cutting the ellipse in Q, R; and the tangents at Q, R intersect (on SP) in T. Shew that a conic can be described with focus S, and directrix PZ, to pass through Q, R and T; and that TZ will be the tangent at T.

52. TP, TQ are tangents to an ellipse at P and Q; one circle touches TP at P and meets TQ in Q and Q'; another touches TQ at Q and meets TP in P and P'; prove that PQ' and QP' are divided in the same ratio by the ellipse.

53. If a chord $RPQV$ meet the directrices of an ellipse in R and V, and the circumference in P and Q, then RP and QV subtend, each at the focus nearer to it, angles of which the sum is equal to the angle between the tangents at P and Q.

54. Two tangents are drawn to the same branch of a rectangular hyperbola from an external point; prove that the angles which these tangents subtend at the centre are respectively equal to the angles which they make with the chord of contact.

55. If the normal at a point P of an hyperbola meet the minor axis in g, Pg will be to Sg in a constant ratio.

56. An ordinate NP of an ellipse is produced to meet the auxiliary circle in Q, and normals to the ellipse and circle at P and Q meet in R; RK, RL are drawn perpendicular to the axes; prove that KPL is a straight line, and also that $KP = BC$ and $LP = AC$.

57. If the tangent at any point P cut the axes of a conic, produced if necessary, in T and T', and if C be the centre of the curve, prove that the area of the triangle TCT' varies inversely as the area of the triangle PCN, where PN is the ordinate of P.

58. The circle of curvature of an ellipse at P passes through the focus S, SM is drawn parallel to the tangent at P to meet the diameter PCP' in M; shew that it divides this diameter in the ratio of $3 : 1$.

59. Prove the following construction for a pair of tangents from any external point T to an ellipse of which the centre is C: join CT, let $TPCP'T$ a similar and similarly situated ellipse be drawn, of which CT is a diameter, and P, P' are its points of intersection with the given ellipse; TP, TP' will be tangents to the given ellipse.

60. Through a fixed point a pair of chords of a circle are drawn at right angles: prove that each side of the quadrilateral formed by joining their extremities envelopes a conic of which the fixed point and the centre of the circle are foci.

61. Any conic passing through the four points of intersection of two rectangular hyperbolas will be itself a rectangular hyperbola.

62. R is the middle point of a chord PQ of a rectangular hyperbola whose centre is C. Through R, RQ', RP' are drawn parallel to the tangents at P and Q respectively, meeting CQ, CP in Q', P'. Prove that C, P', R, Q' are concyclic.

63. The tangents at two points Q, Q' of a parabola meet the tangent at P in R, R' respectively, and the diameter through their point of intersection T meets it in K; prove that $PR = KR'$, and that, if QM, $Q'M'$, TN be the ordinates of Q, Q', T respectively to the diameter through P, PN is a mean proportional between PM and PM'.

64. Common tangents are drawn to two parabolas, which have a common directrix, and intersect in P, Q: prove that the chords joining the points of contact in each parabola are parallel to PQ, and the part of each tangent between its points of contact with the two curves is bisected by PQ produced.

65. An ellipse has its centre on a given hyperbola and touches the asymptotes. The area of the ellipse being always a maximum, prove that its chord of contact with the asymptotes always touches a similar hyperbola.

66. A circle and parabola have the same vertex A and a common axis. $BA'C$ is the double ordinate of the parabola which touches the circle at A', the other

extremity of the diameter which passes through A; PP' is any other ordinate of the parabola parallel to this, meeting the axis in N and the chord AB produced in R: shew that the rectangle between RP and RP' is proportional to the square on the tangent drawn from N to the circle.

67. Tangents are drawn at two points, P, P' on an ellipse. If any tangent be drawn meeting those at P, P' in R, R', shew that the line bisecting the angle RSR' intersects RR' on a fixed tangent to the ellipse. Find the point of contact of this tangent.

68. Having given a pair of conjugate diameters of an ellipse, PCP', DCD', let PF be the perpendicular from P on CD, in PF take PE equal to CD, bisect CE in O, and on CE as diameter describe a circle; prove that PO will meet the circle in two points Q and R such that CQ, CR are the directions of the semi-axes, and PQ, PR their lengths.

69. A straight line is drawn through the angular point A of a triangle ABC to meet the opposite side in a; two points O, O' are taken on Aa, and CO, CO' meet AB in c and c', and BO, BO' meet CA in b, b'; shew that a conic passing through $abb'cc'$ will be touched by BC.

70. If TP, TQ are two tangents to a parabola, and any other tangent meets them in Q and R, the middle point of QR describes a straight line.

71. Lines from the centre to the points of contact of two parallel tangents to a rectangular hyperbola and concentric circle make equal angles with either axis of the hyperbola.

72. A line moves between two lines at right angles so as to subtend a right angle and a half at a fixed point on the bisector of the right angle; prove that it touches a rectangular hyperbola.

73. Two cones, whose vertical angles are supplementary, are placed with their vertices coincident and their axes at right angles, and are cut by a plane perpendicular to a common generating line; prove that the directrices of the section of one cone pass through the foci of the section of the other.

74. The normal at a point P of an ellipse meets the curve again in P', and through O, the centre of curvature at P, the chord QOQ' is drawn at right angles to PP'; prove that

$$QO \,.\, OQ' : PO \,.\, OP' :: 2 \,.\, PO : PP'.$$

75. From an external point T, tangents are drawn to an ellipse, the points of contact being on the same side of the major axis. If the focal distances of these points intersect in M and N, TM, TN are tangents to a confocal hyperbola, which passes through M and N.

76. Two tangents to an hyperbola from T meet the directrix in F and F'; prove that the circle, centre T, which touches SF, SF', meets the directrix in two points the radii to which from the point T are parallel to the asymptotes.

77. QR, touching the ellipse at P, is one side of the parallelogram formed by tangents at the ends of conjugate diameters; if the normal at P meet the axes in G and g, prove that QG and Rg are at right angles.

78. If PP' be a double ordinate of an ellipse, and if the normal at P meet CP' in O, prove that the locus of O is a similar ellipse, and that its axis is to the axis of the given ellipse in the ratio

$$AC^2 - BC^2 : AC^2 + BC^2.$$

79. A chord of a conic whose pole is T meets the directrices in R and R'; if SR and $S'R'$ meet in Q, prove that the minor axis bisects TQ.

80. On a parabola, whose focus is S, three points Q, P, Q' are taken such that the angles PSQ, PSQ' are equal; the tangent at P meets the tangents at Q, Q' in T, T': shew that

$$TQ : T'Q' :: SQ : SQ'.$$

81. If from any point P of a parabola perpendiculars PN, PL are let fall on the axis and the tangent at the vertex, the line LN always touches another parabola.

82. PQ is any diameter of a section of a cone whose vertex is V; prove that $VP + VQ$ is constant.

83. If SY, SK are the perpendiculars from a focus on the tangent and normal at any point of a conic, the straight line YK passes through the centre of the conic.

84. If the axes of two parabolas are in the same direction, their common chord bisects their common tangents.

85. Find the position of the normal chord which cuts off from a parabola the least segment.

86. From the point in which the tangent at any point P of an hyperbola meets either asymptote perpendiculars PM, PN are let fall upon the axes. Prove that MN passes through P.

87. If two parabolas whose latera recta have a constant ratio, and whose foci are two given points S, S', have a contact of the second order at P, the locus of P is a circle.

88. Find the class of plane curves such that, if from a fixed point in the plane, perpendiculars are let fall on the tangent and normal at any point of any one of the curves, the join of the feet of the perpendiculars will pass through another fixed point.

89. If two ellipses have one common focus S and equal major axes, and if one ellipse revolves in its own plane about S, the chord of intersection envelopes a conic confocal with the fixed ellipse.

90. The tangent at any point P of an ellipse meets the axis minor in T and the focal distances SP, HP meet it in R, r. Also ST, HT, produced if necessary, meet the normal at P in Q, q, respectively. Prove that Qr and qR are parallel to the axis major.

91. Two points describe the circumference of an ellipse, with velocities which are to one another in the ratio of the squares on the diameters parallel to their respective directions of motion. Prove that the locus of the point of intersection of their directions of motion will be an ellipse, confocal with the given one.

92. If AA' be the axis major of an elliptic section of a cone, vertex O, and if AG, $A'G'$ perpendicular to AV, $A'V$ meet the axis of the cone in G and G', and GU, $G'U'$ be the perpendiculars let fall on AA', prove that U and U' are the centres of curvature at A and A'.

93. By help of the geometry of the cone, or otherwise, prove that the sum of the tangents from any point of an ellipse to the circles of curvature at the vertices is constant.

94. If two tangents be drawn to a section of a cone, and from their intersection two straight lines be drawn to the points where the tangent plane to the cone through one of the tangents touches the focal spheres, prove that the angle contained by these lines is equal to the angle between the tangents.

95. If CP, CD are conjugate semi-diameters and if through C is drawn a line parallel to either focal distance of P, the perpendicular from D upon this line will be equal to half the minor axis.

96. The area of the parallelogram formed by the tangents at the ends of any pair of diameters of a central conic varies inversely as the area of the parallelogram formed by joining the points of contact.

97. Shew how to draw through a given point a plane which will have the given point for (1) focus, (2) centre, of the section it makes of a given right circular cone: noticing any limitations in the position of the point which may be necessary.

98. In the first figure of Art. 148, if a plane be drawn intersecting the focal spheres in two circles and the cone in an ellipse, the sum or difference of the tangents from any point of the ellipse to the circles is constant.

99. If sections of a right cone be made, perpendicular to a given plane, such that the distance between a focus of a section and that vertex which lies on one of the generating lines in the given plane be constant, prove that the transverse axes, produced if necessary, of all sections will touch one of two fixed circles.

100. A sphere rolls in contact with two intersecting straight wires; prove that its centre describes an ellipse.

CHAPTER XI.

HARMONIC PROPERTIES, POLES AND POLARS.

198. DEF. *A straight line is harmonically divided in two points when the whole line is to one of the extreme parts as the other extreme part is to the middle part.*

Thus AD is harmonically divided in C and B, when

$$AD : AC :: BD : BC.$$

A C B D

This definition may also be presented in the following form.

The straight line AB is harmonically divided in C and D, when it is divided internally in C, and externally in D, in the same ratio.

Under these circumstances the four points A, C, B, D constitute *an Harmonic Range*, and if through any point O four straight lines OA, OC, OB, OD be drawn, these four lines constitute an *Harmonic Pencil.*

PROP. I. *If a straight line be drawn parallel to one of the rays of an harmonic pencil, its segments made by the other three will be equal, and any straight line is divided harmonically by the four rays.*

Let $ACBD$ be the given harmonic range, and draw ECF through C parallel to OD, and meeting OA, OB in E and F.

Then $$AD : AC :: OD : EC,$$
and $$BD : BC :: OD : CF;$$
but from the definition

$$AD : AC :: BD : BC;$$

$$\therefore EC = CF,$$

and any other line parallel to ECF is obviously bisected by OC.

Next, let $acbd$ be any straight line cutting the pencil, and draw ecf parallel to Od; so that $ec = cf$.

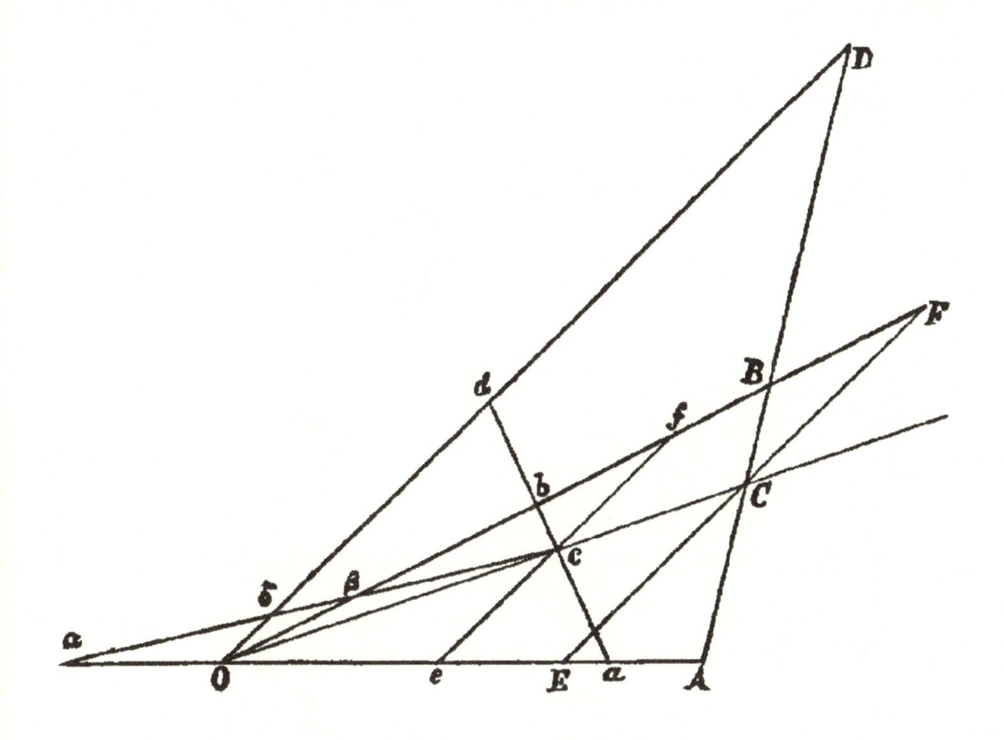

Then $$ad : ac :: Od : ec,$$

and $$bd : bc :: Od : cf;$$

$$\therefore ad : ac :: bd : bc;$$

that is, *acbd* is harmonically divided.

If the line $c\beta\delta\alpha$ be drawn cutting AO produced,

then $$\alpha\delta : \alpha c :: O\delta : ec,$$

and $$\beta\delta : \beta c :: O\delta : cf;$$

$$\therefore \alpha\delta : \alpha c :: \beta\delta : \beta c,$$

or $$\alpha c : \alpha\delta :: \beta c : \beta\delta,$$

and similarly it may be shewn in all other cases that the line is harmonically divided.

199. PROP. II. *The pencil formed by two straight lines and the bisectors of the angles between them is an harmonic pencil.*

For, if OA, OB be the lines, and OC, OD the bisectors, draw KPL parallel to OC and meeting OA, OD, OB. Then the angles OKL, OLK are obviously equal, and the angles at P are right angles; therefore $KP = PL$, and the pencil is harmonic.

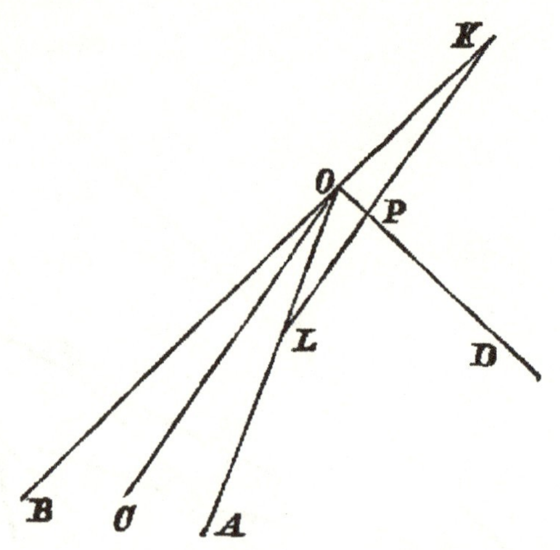

200. Prop. III. *If $ACBD$, $Acbd$ be harmonic ranges, the straight lines Cc, Bb, Dd will meet in a point, as also Cd, cD, Bb.*

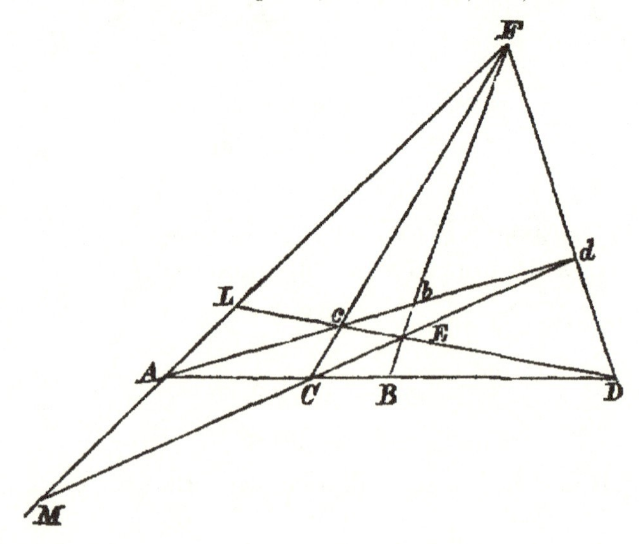

For, if Cc, Dd meet in F, join Fb; then the pencil $F(Acbd)$ is harmonic, and will be cut harmonically by AD.

Hence Fb produced will pass through B.

Similarly, if Cd, cD meet in E,
$E(Acbd)$ is harmonic, and therefore bE produced will pass through B.

Harmonic Properties of a Quadrilateral.

In the preceding figure, let $CcdD$ be any quadrilateral; and let dc, DC meet in A, Cd, cD in E, and Cc, Dd in F.

Then taking b and B so as to divide Acd and ACD harmonically, the ranges $Acbd$ and $ACBD$ are harmonic, and therefore Bb passes through both E and F.

Similarly it can be shewn that AF is divided harmonically in L and M, by Dc and dC.

For $E(Acbd)$ is harmonic and therefore the transversal $ALFM$ is harmonically divided.

201. PROP. IV. *If $ACBD$ be an harmonic range, and E the middle point of CD,*

$$EA \,.\, EB = EC^2.$$

A C B E D

For
$$AD : AC :: BD : BC,$$
or
$$AE + EC : AE - EC :: EC + EB : EC - EB;$$
$$\therefore AE : EC :: EC : EB,$$
or
$$AE \,.\, EB = EC^2 = ED^2.$$

Hence also, conversely, if $EC^2 = ED^2 = AE \,.\, EB$, the range $ACBD$ is harmonic, C and D being on opposite sides of E.

Hence, if a series of points A, a, B, b, ... on a straight line be such that

$$EA \,.\, Ea = EB \,.\, Eb = EC \,.\, Ec \ldots$$
$$= EP^2,$$

and if $EQ = EP$, then the several ranges $(APaQ)$, $(BPbQ)$, &c. are harmonic.

202. DEF. A system of pairs of points on a straight line such that

$$EA \,.\, Ea = EB \,.\, EB \,.\, Eb = \ldots \quad = EP^2 = EQ^2$$

is called a system in *Involution*, the point E being called the centre and P, Q the foci of the system.

Any two corresponding points A, a, are called *conjugate* points, and it appears from above that any two conjugate points form, with the foci of the system, an harmonic range.

It will be noticed that a focus is a point at which conjugate points coincide, and that the existence of a focus is only possible when the points A and a are both on the same side of the centre.

203. PROP. V. *Having given two pairs of points, A and a, B and b, it is required to find the centre and foci of the involution.*

If E be the centre,

$$EA : EB :: Eb : Ea;$$
$$\therefore EA : AB :: Eb : ab,$$
or
$$EA : Eb :: AB : ab.$$

This determines E, and the foci P and Q are given by the relations
$$EP^2 = EQ^2 = EA \,.\, EA.$$

We shall however find the following relation useful.

Since
$$EA : Eb :: EB : Ea;$$
$$\therefore EA : Ab :: EB : aB,$$
or
$$EA : EB :: Ab : aB;$$
but
$$Eb : EA :: ab : AB;$$
$$\therefore Eb : EB :: Ab \,.\, ba : AB \,.\, Ba.$$

Again,
$$Qb : Pb :: QB : PB;$$
$$\therefore Qb - Pb : Pb :: QB - PB : PB,$$
$$2 \,.\, EP : Pb :: 2 \,.\, EB : BP;$$
$$\therefore Pb^2 : PB^2 :: EP^2 : EB^2,$$
$$:: Eb : EB$$
$$:: Ab \,.\, ba : AB \,.\, Ba.$$

This determines the ratio in which Bb is divided by P.

204. If $QAPa$ be an harmonic range and E the middle point of PQ, and if a circle be described on PQ as diameter, the lines joining any point R on this circle with P and Q will bisect the angles between AR and aR.

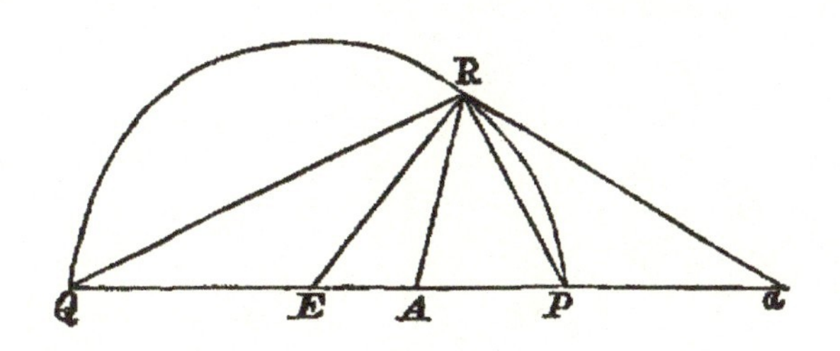

For $$EA \,.\, Ea = EP^2 = ER^2;$$
$$\therefore EA : ER :: ER : Ea,$$
and the triangles ARE, aRE are similar.

Hence $$AR : aR :: EA : ER$$
$$:: EA : EP.$$
But $$Ea : EP :: EP : EA;$$
$$\therefore aP : EP :: AP : EA.$$
Hence $$AR : aR :: AP : aP,$$
and ARa is bisected by RP.

Hence, if A and a, B and b be conjugate points of a system in involution of which P and Q are the foci, it follows that AB and ab subtend equal angles at any point of the circle on PQ as diameter.

This fact also affords a means of obtaining the relations of Art. 203.

We must observe that if the points A, a are on one side of the centre and B, b on the other, the angles subtended by AB, ab are supplementary to each other.

205. PROP. VI. *If four points form an harmonic range, their conjugates also form an harmonic range.*

Let A, B, C, D be the four points, a, b, c, d their conjugates.

Then, as in the eighth line of Art. 203,
$$EA : Ed :: AD : ad,$$
or $$ED : Ea :: AD : ad;$$
$$\therefore AD \,.\, Ea = ED \,.\, ad.$$

Similarly $$AC \,.\, Ea = EC \,.\, ac,$$
$$BD \,.\, Eb = ED \,.\, bd,$$
$$BC \,.\, Eb = EC \,.\, bc.$$

But, $ABCD$ being harmonic,
$$AD : AC :: BD : BC;$$
$$\therefore ED \,.\, ad : EC \,.\, ac :: ED \,.\, bd : EC \,.\, bc.$$

Hence $$ad : ac :: bd : bc,$$
or the range of the conjugates is harmonic.

206. PROP. VII. *If a system of conics pass through four given points, any straight line will be cut by the system in a series of points in involution.*

The four fixed points being C, D, E, F, let the line meet one of the conics in A and a, and the straight lines CF, ED, in B and b.

Then the rectangles $AB \,.\, Ba$, $CB \,.\, BF$ are in the ratio of the squares on parallel diameters, as also are $Ab \,.\, ba$ and $Db \,.\, bE$.

But the squares on the diameters parallel to CF, ED are in the constant ratio $KF \,.\, KC : KE \,.\, KD$; and, the line Bb being given in position, the rectangles $CB \,.\, BF$ and $Db \,.\, bE$ are given; therefore the rectangles $AB \,.\, Ba$, $Ab \,.\, ba$ are in a constant ratio.

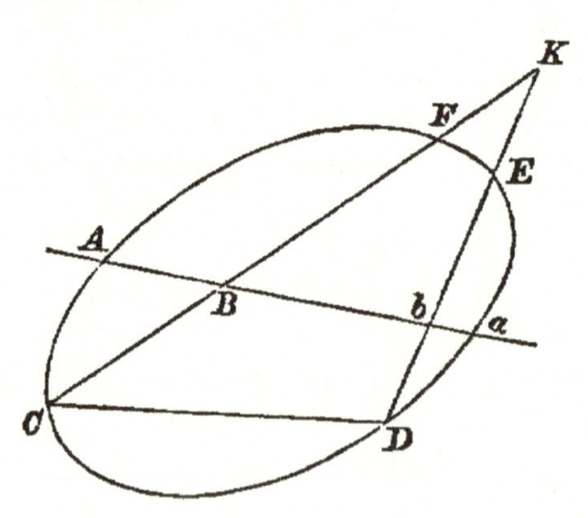

But (Art. 203) this ratio is the same as that of PB^2 to Pb^2, if P be a focus of the involution A, a, B, b.

Hence P is determined, and all the conics cut the line Bb in points which form with B, b a system in involution.

We may observe that the foci are the points of contact of the two conics which can be drawn through the four points touching the line, and that the centre is the intersection of the line with the conic which has one of its asymptotes parallel to the line.

207. PROP. VIII. *If through any point two tangents be drawn to a conic, any other straight line through the point will be divided harmonically by the curve and the chord of contact.*

Let AB, AC be the tangents, $ADFE$ the straight line.

Through D and E draw $GDHK$, $LEMN$ parallel to BC.

Then the diameter through A bisects DH, and BC, and therefore bisects GK; hence $GD = HK$, and similarly $LE = MN$.

Also $$LE : EN :: GD : DK;$$
$$\therefore LE \,.\, EN : LE^2 :: GD \,.\, DK : GD^2,$$
or $$LE \,.\, LM : GD \,.\, GH :: LE^2 : GD^2$$
$$:: LA^2 : GA^2.$$

But $$LE \,.\, LM : GD \,.\, GH :: LB^2 : BG^2;$$
hence $$AL : AG :: BL : BG,$$

and therefore $$AE : AD :: FE : FD,$$
that is, $ADFE$ is harmonically divided.

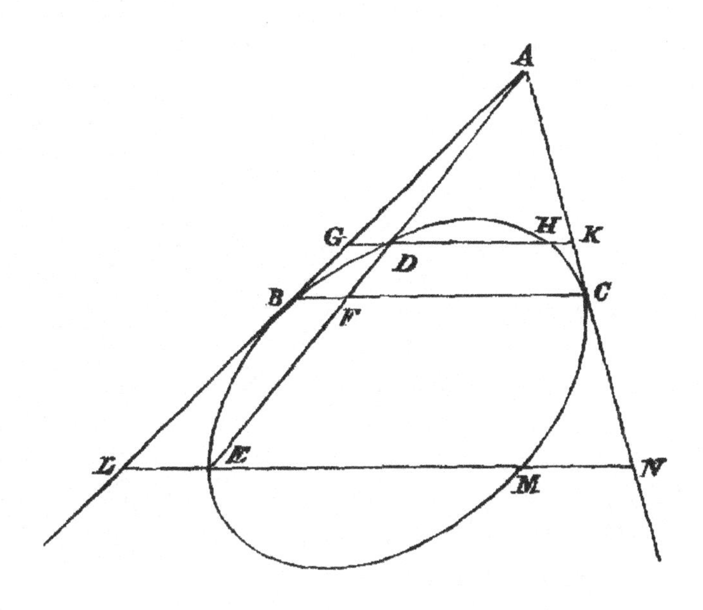

208. PROP. IX. *If two tangents be drawn to a conic, any third tangent is harmonically divided by the two tangents, the curve, and the chord of contact.*

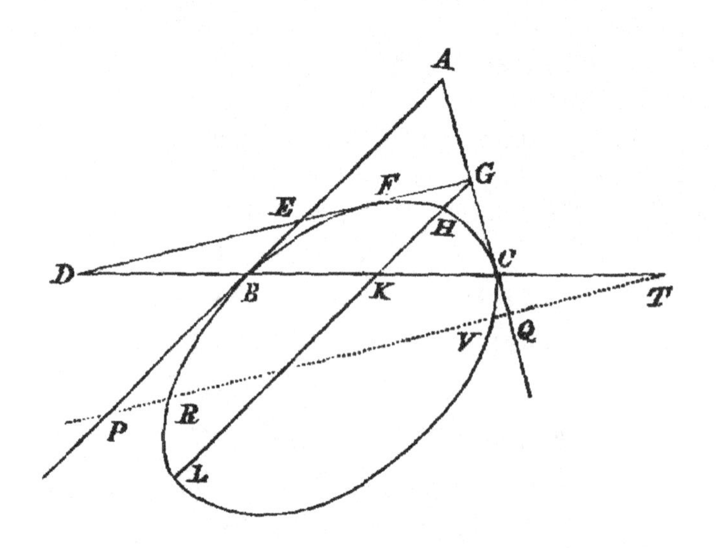

Let $DEFG$ be the third tangent, and through G, the point in which it meets AC, draw $GHKL$ parallel to AB, cutting the curve and the chord of contact in H, K, L.

Then $$GH\,.\,GL : GC^2 :: AB^2 : AC^2$$
$$:: GK^2 : GC^2;$$
$$\therefore GH\,.\,GL = GK^2.$$

Hence $$DG^2 : DE^2 :: GK^2 : EB^2$$
$$:: GH\,.\,GL : EB^2$$
$$:: FG^2 : FE^2;$$

that is, $DEFG$ is an harmonic range.

209. PROP. X. *If any straight line meet two tangents to a conic in P and Q, the chord of contact in T and the conic in R and V,*

$$PR\,.\,PV : QR\,.\,QV :: PT^2 : QT^2.$$

Taking the preceding figure, draw the tangent $DEFG$ parallel to PQ.

Then $$PR \,.\, PV : EF^2 :: PB^2 : BE^2$$
$$:: PT^2 : DE^2;$$

and $$QR \,.\, QV : GF^2 :: QC^2 : GC^2$$
$$:: QT^2 : DG^2;$$

but $$EF : DE :: GF : DG;$$
$$\therefore PR \,.\, PV : PT^2 :: QR \,.\, QV : QT^2.$$

210. PROP. XI. *If chords of a conic be drawn through a fixed point the pairs of tangents at their extremities will intersect in a fixed line.*

Let B be the fixed point and C the centre, and let CB meet the curve in P.

Take A in CP such that

$$CA : CP :: CP : CB;$$

then B is the middle point of the chord of contact of the tangents AQ, AR.

Draw any chord EBF, and let the tangents at E and F meet in G: also join CG and draw PN parallel to EF.

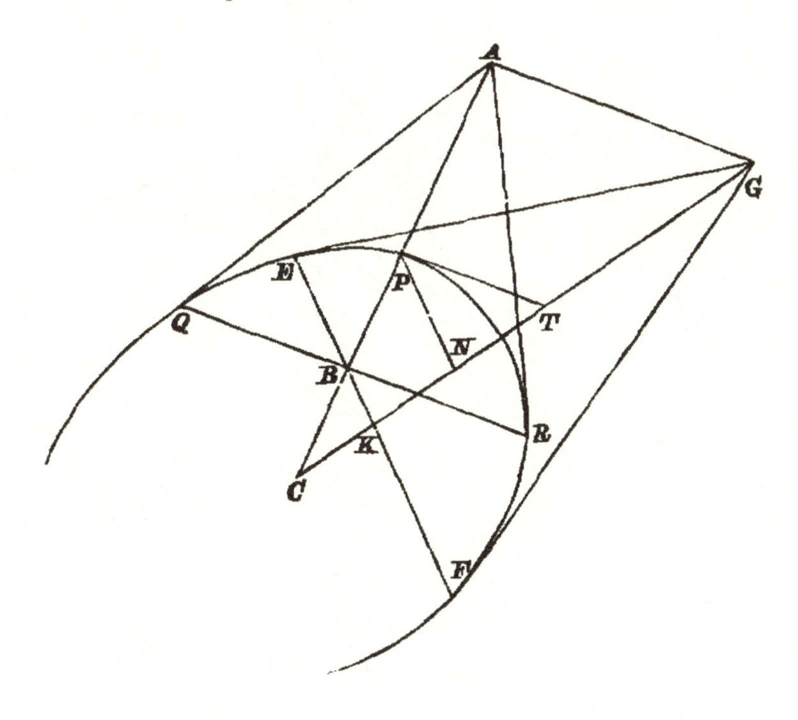

Then if CG meet EF in K and the tangent at P in T,

$$CK \,.\, CG = CN \,.\, CT;$$
$$\therefore CG : CT :: CN : CK$$
$$:: CP : CB$$
$$:: CA : CP;$$

hence AG is parallel to PT, and the point G therefore lies on a fixed line.

If the conic be a parabola, we must take AP equal to BP: then, remembering that KG and NT are bisected by the curve, the proof is the same as before.

211. If A be the fixed point, let CA meet the curve in P, and take B in CP such that

$$CB : CP :: CP : CA;$$

then B is the middle point of the chord of contact of the tangents AQ, AR. Draw any chord AEF, and let the tangents at E and F meet in G; also join CG and draw PN parallel to EF.

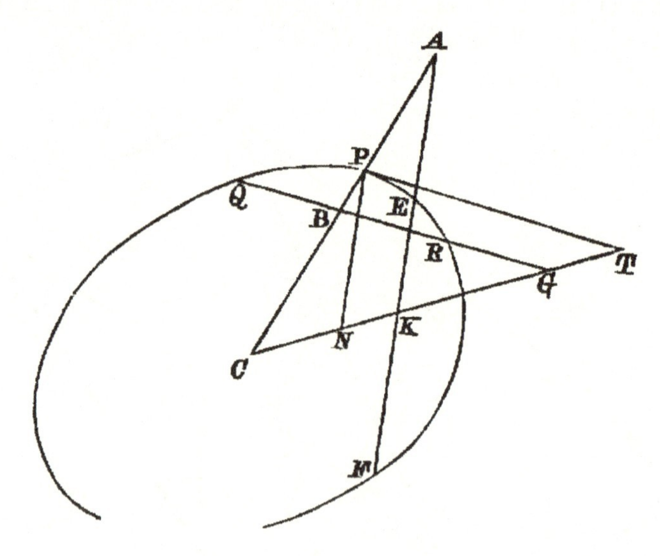

Then $$CK \,.\, CG = CN \,.\, CT;$$
$$\therefore CG : CT :: CN : CK$$
$$:: CP : CA$$
$$:: CB : CP;$$

$\therefore BG$ is parallel to PT and coincides with the chord of contact QR.

Hence, conversely, if from points on a straight line pairs of tangents be drawn to a conic, the chords of contact will pass through a fixed point.

Poles and Polars.

212. DEF. The straight line which is the locus of the points of intersection of tangents at the extremities of chords through a fixed point is called the *polar* of the point.

Also, if from points in a straight line pairs of tangents be drawn to a conic, the point in which all the chords of contact intersect is called the *pole* of the line.

If the pole be without the curve the polar is the chord of contact of tangents from the pole.

If the pole be on the curve the polar is the tangent at the point.

It follows at once from these definitions that the focus of a conic is the pole of the directrix, and that the foot of the directrix is the pole of the latus rectum.

213. PROP. XII. *A straight line drawn through any point is divided harmonically by the point, the curve, and the polar of the point.*

If the point be without the conic this is already proved in Art. 207.

If it be within the conic, as B in the figure of Art. 210, then, drawing any chord $FBEV$ meeting in V the polar of B, which is AG, the chord of contact of tangents from V passes through B, by Art. 211, and the line $VEBF$ is therefore harmonically divided.

Hence the polar may be constructed by drawing two chords through the pole and dividing them harmonically; the line joining the points of division is the polar.

Or, in the figure of Art. 210,

$$CB \,.\, CA = CP^2,$$

so that the polar of B is obtained by taking the point A on the diameter through B, at the distance from C given by the above relation, and then drawing AG parallel to the diameter which is conjugate to CP.

COR. Hence it follows that *the centre of a conic is the pole of a line at an infinite distance.*

For, if CB is diminished indefinitely, CA is increased indefinitely.

214. PROP. XIII. *The polars of two points intersect in the pole of the line joining the two points.*

For, if A, B be the two points and O the pole of AB, the line AO is divided harmonically by the curve, and therefore the polar of A passes through the point O.

Similarly the polar of B passes through O;

That is, the polars of A and B intersect in the pole of AB.

215. PROP. XIV. *If a quadrilateral be inscribed in a conic, its opposite sides and diagonals will intersect in three points such that each is the pole of the line joining the other two.*

Let $ABCD$ be the quadrilateral, F and G the points of intersection of AD, BC, and of DC, AB.

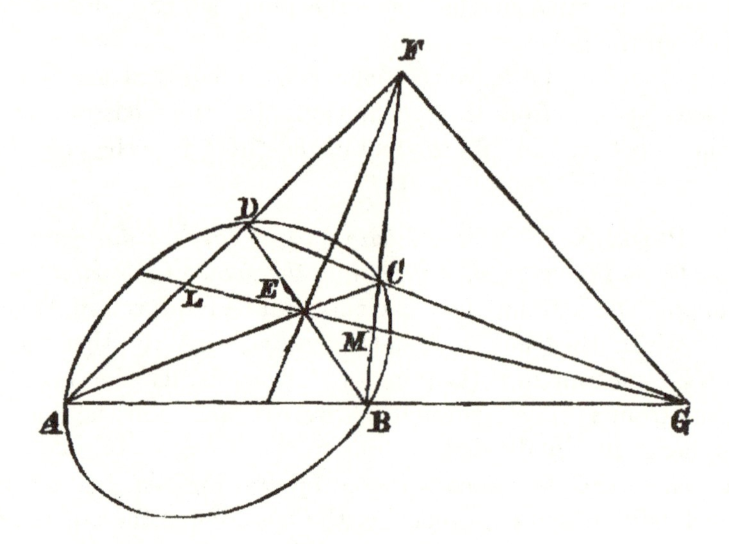

Let EG meet FA, FB, in L and M.

Then (Art. 200) $FDLA$ and $FCMB$ are harmonic ranges;

Therefore L and M are both on the polar of F (Art. 213), and EG is the polar of F.

Similarly, EF is the polar of G, and therefore E is the pole of FG (Art. 214).

216. DEF. If each of the sides of a triangle be the polar, with regard to a conic, of the opposite angular point, the triangle is said to be *self-conjugate* with regard to the conic.

Thus the triangle EGF in the above figure is self-conjugate.

To construct a self-conjugate triangle, take a straight line AB and find its pole C.

Draw through C any straight line CD cutting AB in D, and find the pole E of CD, which lies on AB: then CDE is self-conjugate.

217. Prop. XV. *If a quadrilateral circumscribe a conic, its three diagonals form a self-conjugate triangle.*

Let the polar of F (that is, the chord of contact $P'P$), meet FG in R; then, since R is on the polar of F, it follows that F is on the polar of R.

Now $F(AEBG)$ is harmonic (Art. 200), and, if FE meet $P'P$ in T, $P'TPR$ is an harmonic range; hence, by the theorem of Art. 213, FT, *i.e.* FE, is the polar of R.

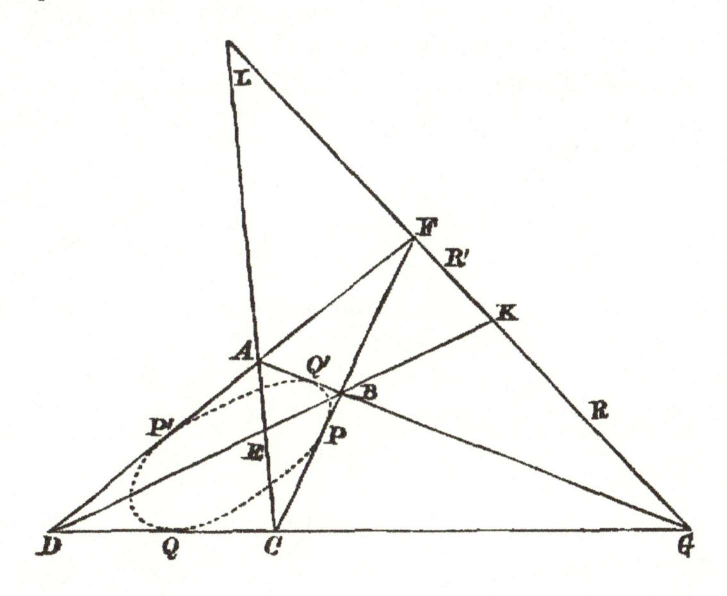

Similarly, if the other chord of contact QQ' meet FG in R', GE is the polar of R';

$$\therefore E \text{ is the pole of } RR', \text{ that is, of } LK.$$

Again, $DEBK$ is harmonic, and therefore the pencil $C(QEPK)$ is harmonic.

Hence, if QP meet AC in S and CK in V, $QSPV$ is harmonic, and therefore S is on the polar of V.

But S is on the polar of C; therefore CV, that is, CK, is the polar of S.

Similarly, if $P'Q'$ meet AC in S', AK is the polar of S'.

Hence it follows that K is the pole of SS', that is, of EL; ELK is therefore a self-conjugate triangle.

218. PROP. XVI. *If a system of conics have a common self-conjugate triangle, any straight line passing through one of the angular points of the triangle is cut in a series of points in involution.*

For, if ABC be the triangle, and a line $APDQ$ meet BC in D, and the conic in P and Q, $APDQ$ is an harmonic range, and all the pairs of points P, Q form with A and D an harmonic range.

Hence the pairs of points form a system in involution, of which A and D are the foci.

219. PROP. XVII. *The pencil formed by the polars of the four points of an harmonic range is an harmonic pencil.*

Let $ABCD$ be the range, O the pole of AD.

Let the polars Oa, Ob, Oc, Od meet AD in a, b, c, d, and let AD meet the conic in P and Q.

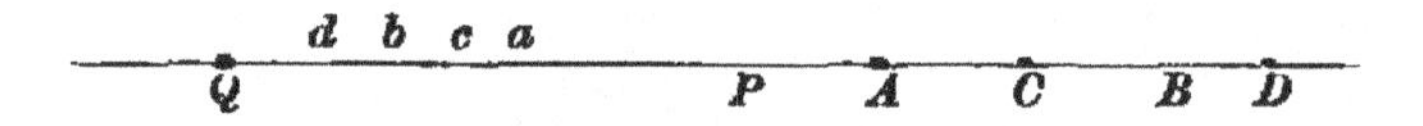

Then $APaQ$, $CPcQ$, &c. are harmonic ranges; and therefore (Arts. 201, 202) a, c, b, d are the conjugates of A, C, B, D.

Hence (Art. 205) the range $acbd$ is harmonic, and therefore the pencil O $(acbd)$ is harmonic.

EXAMPLES.

1. If PSP' is a focal chord of a conic, any other chord through S is divided harmonically by the directrix and the tangents at P and P'.

2. If two sections of a right cone be taken, having the same directrix, the straight line joining the corresponding foci will pass through the vertex.

3. If a series of circles pass through the same two points, any transversal will be cut by the circles in a series of points in involution.

4. If O be the centre of the circle circumscribing a triangle ABC, and $B'C'$, $C'A'$, $A'B'$, the respective polars with regard to a concentric circle of the points A, B, C, prove that O is the centre of the circle inscribed in the triangle $A'B'C'$.

5. OA, OB, OC being three straight lines given in position, shew that there are three other straight lines each of which forms with OA, OB, OC an harmonic pencil; and that each of the three OA, OB, OC forms with the second three an harmonic pencil.

6. The straight line $ACBD$ is divided harmonically in the points C, B; prove that if a circle be described on CD as diameter, any circle passing through A and B will cut it at right angles.

7. Three straight lines AD, AE, AF are drawn through a fixed point A, and fixed points C, B, D are taken in AD, such that $ACBD$ is an harmonic range. Any straight line through B intersects AE and AF in E and F, and CE, DF intersect in P; DE, CF in Q. Shew that P and Q always lie in a straight line through A, forming with AD, AE, AF an harmonic pencil.

8. CA, CB are two tangents to a conic section, O a fixed point in AB, POQ any chord of the conic; prove that the intersections of AP, BQ, and also of AQ, BP lie in a fixed straight line which forms with CA, CO, CB an harmonic pencil.

9. If three conics pass through the same four points, the common tangent to two of them is divided harmonically by the third.

10. Two conics intersect in four points, and through the intersection of two of their common chords a tangent is drawn to one of them; prove that it is divided harmonically by the other.

11. Prove that the two tangents through any point to a conic, any line through the point and the line to the pole of the last line, form an harmonic pencil.

12. The locus of the poles, with regard to the auxiliary circle, of the tangents to an ellipse, is a similar ellipse.

13. The asymptotes of an hyperbola and any pair of conjugate diameters form an harmonic pencil.

14. PSQ and $PS'R$ are two focal chords of an ellipse; two other ellipses are described having P for a common focus, and touching the first ellipse at Q and R respectively. The three ellipses have equal major axes. Prove that the directrices of the last two ellipses pass through the pole of QR.

15. Tangents from T touch an ellipse in P and Q, and PQ meets the directrices in R and R'; shew that PR and QR' subtend equal angles at T.

16. The poles of a given straight line, with respect to sections through it of a given cone, all lie upon a straight line passing through the vertex of the cone.

17. If from a given point in the axis of a conic a chord be drawn, the perpendicular from the pole of the chord upon the chord will meet the axis in a fixed point.

18. Q is any point in the tangent at a point P of a conic; QG perpendicular to CP meets the normal at P in G, and QE perpendicular to the polar of Q meets the normal at P in E; prove that EG is constant and equal to the radius of curvature at P.

19. The line joining two fixed points A and B meets the two fixed lines OP, OQ in P and Q.

A conic is described so that OP and OQ are the polars of A and B with respect to it. Prove that the locus of its centre is the line OR, where R divides AB so that

$$AR : RB :: QR : RP.$$

20. If from a point O in the normal at a point R of an ellipse tangents OP, OQ are drawn, the angles PRO, QRO are equal.

21. The focal distances of a point on a conic meet the curve again in Q, R; shew that the pole of QR will lie upon the normal at the first point.

22. The tangent at any point A of a conic is cut by two other tangents and their chord of contact in B, C, D; shew that $(ABDC)$ is harmonic.

23. A rectangular hyperbola circumscribes a triangle ABC; if D, E, F be the feet of the perpendiculars from A, B, C on the opposite sides, the loci of the poles of the sides of the triangle ABC are the lines EF, FD, DE.

24. Two common chords of a given ellipse and a circle pass through a given point; shew that the locus of the centres of all such circles is a straight line through the given point.

25. If $ABCD$ is a quadrilateral inscribed in a conic, and if AD, BC meet in P, and AC, BD in Q, PQ passes through the pole of AB.

26. PCP' is any diameter of an ellipse. The tangents at the points D, E intersect in F, and PE, $P'D$ intersect in G. Shew that FG is parallel to DCD'.

27. PP' is a chord of a conic, QQ' any chord through its pole. Prove that lines drawn from P parallel to the tangents at Q and Q' to meet $P'Q$, and $P'Q'$ respectively are bisected by QQ'.

28. If the pencil joining four fixed points on a conic to any one point on the conic is harmonic, the pencil joining the fixed points to any point on the conic is harmonic.

29. If PQ is the chord of a conic having its pole on the chord AB or AB produced, and if Qq is the chord parallel to AB, then Pq bisects AB.

30. If a quadrilateral circumscribe a conic, the intersection of the lines joining opposite points of contact is the same as the intersection of the diagonals.

CHAPTER XII.

Reciprocal Polars.

220. The pole of a line with regard to any conic being a point and the polar of a point a line, it follows that any system of points and lines can be transformed into a system of lines and points.

This process is called *reciprocation*, and it is clear that any theorem relating to the original system will have its analogue in the system formed by reciprocation.

Thus, if a series of lines be concurrent, the corresponding points are collinear; and the theorem of Art. 219 is an instance of the effect of reciprocation.

221. Def. If a point move in a curve (C), its polar will always touch some other curve (C'); this latter curve is called the reciprocal polar of (C) with regard to the auxiliary conic.

Prop. I. *If a curve C' be the polar of C, then will C be the polar of C'.*

For, if P, P' be two consecutive points of C, the intersection of the polars of P and P' is a point Q, which is the pole of the line PP'.

But the point Q is ultimately, when P and P' coincide, the point of contact of the curve which is touched by the polar of P.

Hence the polar of any point Q of C' is a tangent to the curve C.

222. So far we have considered poles and polars generally with regard to any conic; we shall now consider the case in which a circle is the auxiliary curve.

In this case, if AB be a line, P its pole, and CY the perpendicular from the centre of the circle on AB, the rectangle $CP.\ CY$ is equal to the square on the radius of the circle.

A simple construction is thus given for the pole of a line, or the polar of the point.

As an illustration take the theorem of the existence of the orthocentre in a triangle.

Let AOD, BOE, COF be the perpendiculars, O being the orthocentre.

The polar reciprocal of the line BC is a point A', and of the point A a line $B'C'$.

To the line AD corresponds a point P on $B'C'$, and since ADB is a right angle, it follows that PSA' is a right angle, S being the centre of the auxiliary circle.

And, similarly, if SQ, SR, perpendiculars to SB', SC', meet $C'A'$ and $A'B'$ in Q and R, these points correspond to BE and CF.

But AD, BE, CF are concurrent;

$$\therefore P, Q, R \text{ are collinear.}$$

Hence the reciprocal theorem,

If from any point S lines be drawn perpendicular respectively to SA', SB', SC', and meeting $B'C'$, $C'A'$, $A'B'$ in P, Q, and R, these points are collinear.

As a second illustration take the theorem,

If A, B be two fixed points, and AC, BC at right angles to each other, the locus of C is a circle.

Taking O, the middle point of AB, as the centre of the auxiliary circle, the reciprocals of A and B are two parallel straight lines, PE, QF, perpendicular to AB; the reciprocals of AC, BC are points P, Q on these lines such that POQ is a right angle, and PQ is the reciprocal of C.

Hence, the locus of C being a circle, it follows that PQ always touches a circle.

The reciprocal theorem therefore is,

If a straight line PQ, bounded by two parallel straight lines, subtend a right angle at a point O, halfway between the lines, the line PQ always touches a circle, having O for its centre.

223. PROP. II. *The reciprocal polar of a circle with regard to another circle, called the auxiliary circle, is a conic, a focus of which is the centre of the auxiliary circle, and the corresponding directrix the polar of the centre of the reciprocated circle.*

Let S be the centre of the auxiliary circle, and KX the polar of C, the centre of the reciprocated circle.

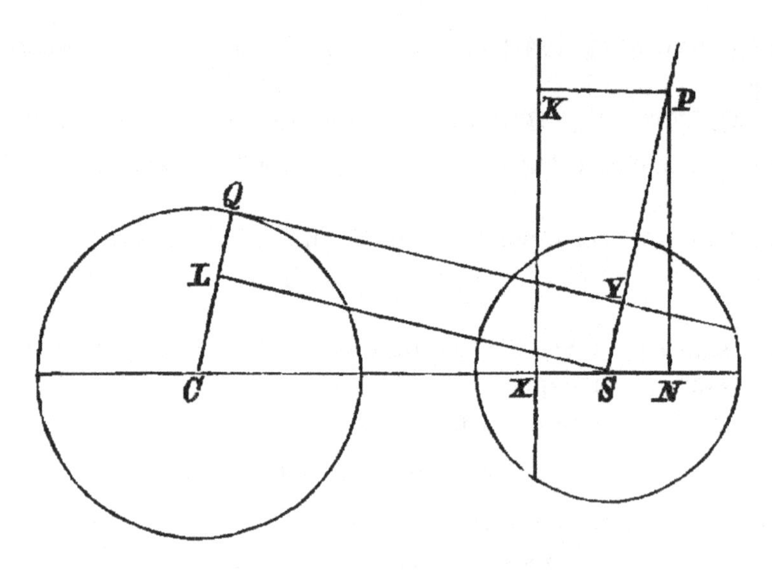

Then, if P be the pole of a tangent QY to the circle C, SP meeting this tangent in Y,

$$SP \,.\, SY = SX \,.\, SC.$$

Therefore, drawing SL parallel to QY,

$$SP : SC :: SX : QL.$$

But, by similar triangles,

$$SP : SC :: SN : CL;$$

$$\therefore SP : SC :: NX : CQ,$$

or

$$SP : PK :: SC : CQ.$$

Hence the locus of P is a conic, focus S, directrix KX and having for its eccentricity the ratio of SC to CQ.

The reciprocal polar of a circle is therefore an ellipse, parabola, or hyperbola, as the point S is within, upon, or without the circumference of the circle.

224. PROP. III. *To find the latus rectum and axes of the reciprocal conic.*

The ends of the latus rectum are the poles of the tangents parallel to SC.

Hence, if SR be the semi-latus rectum,

$$SR \,.\, CQ = SE^2,$$

SE being the radius of the auxiliary circle.

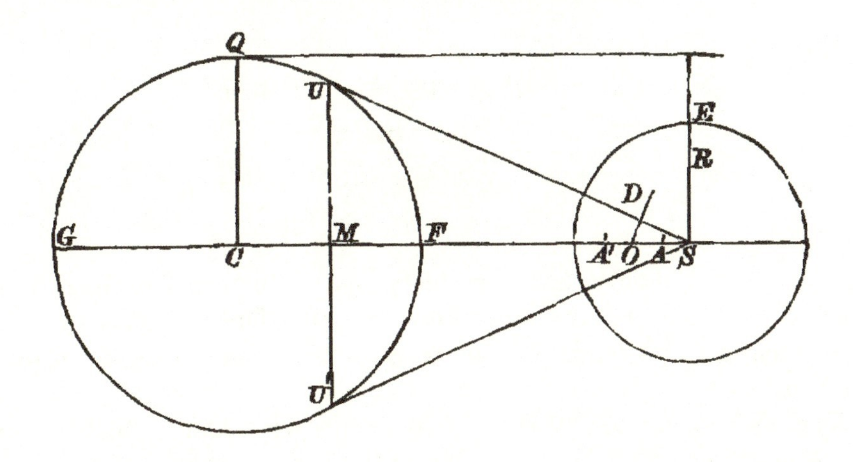

The ends of the transverse axis A, A' are the poles of the tangents at F and G;

$$\therefore SA \,.\, SG = SE^2$$

and $$SA' \,.\, SF = SE^2.$$

Let SU, SU' be the tangents from S, then

$$SG \,.\, SF = SU^2,$$

$$\left.\begin{array}{r} \therefore SA' : SG :: SE^2 : SU^2 \\ SA : SF :: SE^2 : SU^2 \end{array}\right\} \quad (\alpha).$$

(the second line preceded by "and")

Hence $$AA' : FG :: SE^2 : SU^2,$$

or, if O be the centre of the reciprocal,

$$AO : CQ :: SE^2 : SU^2.$$

Again, if BOB' be the conjugate axis,

$$BO^2 = SR \,.\, AO;$$

therefore, since $$SE^2 = SR \,.\, CQ,$$

$$BO^2 : SE^2 :: AO : CQ$$

$$:: SE^2 : SU^2$$

and $$BO \,.\, SU = SE^2.$$

The centre O, it may be remarked, is the pole of UU'.

For, from the relations (α),

$$SE^2 : SU^2 :: SA + SA' : SF + SG$$
$$:: SO : SC$$
$$:: SO \,.\, SM : SC \,.\, SM;$$
$$\therefore SO \,.\, SM = SE^2.$$

225. In the figures drawn in the two preceding articles, the reciprocal conic is an hyperbola; the asymptotes are therefore the lines through O perpendicular to SU and SU', the poles of these lines being at an infinite distance.

The semi-conjugate axis is equal to the perpendicular from the focus on the asymptote (Art. 103), *i.e.* if OD be the asymptote, SD is equal to the semi-conjugate axis.

Further, since OD is perpendicular to SU, and O is the pole of UU', it follows that D is the pole of CU, and that

$$SD \,.\, SU = SE^2,$$

as we have already shewn.

Again, D, being the intersection of the polars of C and U, is the intersection of SU and the directrix.

226. If the point S be within the circle, so that the reciprocal is an ellipse, the axes are given by similar relations.

Through S draw SV perpendicular to FG, and let UMU' be the polar of S with regard to the circle.

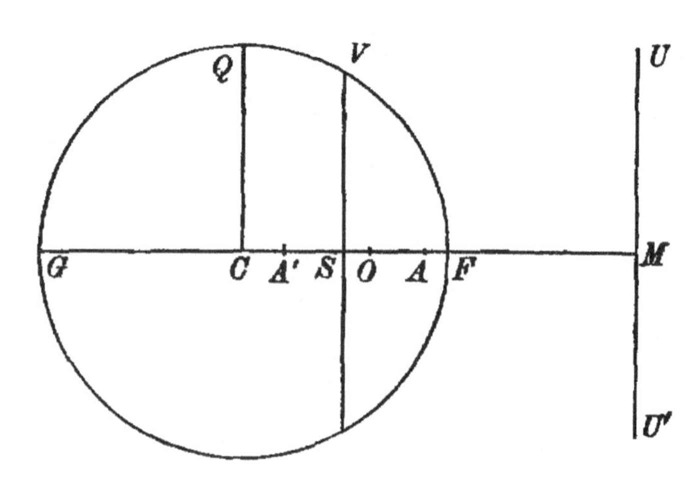

Then $SM \,.\, SC = SC \,.\, CM - SC^2 = CF^2 - SC^2 = SV^2$; also, SE being the radius of the auxiliary circle,

$$SA \,.\, SF = SE^2 = SA' \,.\, SG,$$

and

$$SF \,.\, SG = SV^2;$$

$$\left.\begin{aligned} \therefore SA : SG :: SE^2 : SV^2 \\ SA' : SF :: SE^2 : SV^2 \end{aligned}\right\}.$$

Hence

$$SO : SC :: SE^2 : SV^2,$$

and

$$SO \,.\, SM : SC \,.\, SM :: SE^2 : SV^2;$$

$$\therefore SO \,.\, SM = SE^2,$$

so that O is the pole of UU'.

Again $$SA + SA' : SF + SG :: SE^2 : SV^2,$$

$$\therefore AO : CQ :: SE^2 : SV^2.$$

If RSR' is the latus rectum,

$$SR \,.\, CQ = SE^2,$$

and if BOB' is the minor axis

$$SR \,.\, AO = BO^2;$$

$$\therefore BO^2 : SE^2 :: SE^2 : SV^2,$$

and

$$BO \,.\, SV = SE^2.$$

227. The important Theorem we have just considered enables us to deduce from any property of a circle a corresponding property of a conic, and we are thus furnished with a method, which may serve to give easy proofs of known properties, or to reveal new properties of conics.

In the process of reciprocation we observe that points become lines and lines points; that a tangent to a curve reciprocates into a point on the reciprocal, that a curve inscribed in a triangle becomes a curve circumscribing a triangle, and that when the auxiliary curve is a circle, the reciprocal of a circle is a conic, the latus rectum of which varies inversely as the radius of the circle.

Also, conversely, the reciprocal of a conic with regard to a circle having its centre at a focus of the conic is a circle the centre of which is the reciprocal of the directrix of the conic.

For an ellipse the centre of reciprocation is within the circle, for a parabola it is upon the circle, and for an hyperbola it is outside the circle.

228. We give some transformations of theorems as illustrations of the preceding articles.

THEOREM.	RECIPROCAL.
The line joining the points of contact of parallel tangents of a circle passes through the centre.	The tangents at the ends of a focal chord intersect in the directrix.
The angles in the same segment of a circle are equal.	If a moveable tangent of a conic meet two fixed tangents, the intercepted portion subtends a constant angle at the focus.
Two of the common tangents of two equal circles are parallel.	If two conics have the same focus, and equal latera recta, the straight line joining two of their common points passes through the focus.
The tangent at any point of a circle is perpendicular to the diameter through the point.	The portion of the tangent to a conic between the point of contact and the directrix subtends a right angle at the focus.
A chord of a circle is equally inclined to the tangents at its ends.	The tangents drawn from any point to a conic subtend equal angles at a focus.
If a chord of a circle subtend a constant angle at a fixed point on the curve, the chord always touches a circle.	If two tangents of a conic move so that the intercepted portion of a fixed tangent subtends a constant angle at the focus, the locus of the intersection of the moving tangents is a conic having the same focus and directrix.
If a chord of a circle pass through a fixed point, the rectangle contained by the segments is constant.	The rectangle contained by the perpendiculars from the focus on two parallel tangents is constant.
If two chords be drawn from a fixed point on a circle at right angles to each other, the line joining their ends passes through the centre.	If two tangents of a conic move so that the intercepted portion of a fixed tangent subtends a right angle at the focus, the two moveable tangents meet in the directrix.

THEOREM.	RECIPROCAL.
If a circle be inscribed in a triangle, the lines joining the vertices with the points of contact meet in a point.	If a triangle be inscribed in a conic the tangents at the vertices meet the opposite sides in three points lying in a straight line.
The sum of the reciprocals of the radii of the escribed circles of a triangle is equal to the reciprocal of the radius of the inscribed circle.	With a given point as focus, four conics can be drawn circumscribing a triangle, and the latus rectum of one is equal to the sum of the latera recta of the other three.
The common chord of two intersecting circles is perpendicular to the line joining their centres.	If two parabolas have a common focus, the line joining it to the intersection of the directrices is perpendicular to the common tangent.
If circles pass through two fixed points, the locus of their centres is a straight line.	If conics have a fixed focus and a pair of fixed tangents in common, the corresponding directrices all pass through a fixed point.
Two tangents to a conic at right angles to each other intersect on a fixed circle.	Chords of a circle which subtend a right angle at a fixed point all touch a conic of which that point is a focus.

229. PROP. IV. *A system of coaxal circles can be reciprocated into a system of confocal conics.*

Let X be the point at which the radical axis crosses the line of centres, and let E and S be the limiting points of the system.

Then XE is equal to the length of the tangent XD to any one of the circles, and, therefore, if A is the centre of this circle, AD is the tangent at D to the circle whose centre is X and radius XE.

Hence it follows that $AE \,.\, AS = AD^2$, shewing that UU', the polar of S with regard to the circle A, passes through E.

Reciprocating with regard to S, the centre of the reciprocal curve is the pole of UU', and is consequently fixed; and the conics are therefore confocal.

Hence, if we reciprocate with regard to either limiting point, we obtain confocal conics.

In the particular case in which the circles all touch the radical axis, we obtain confocal and co-axial parabolas.

230. PROP. V. *The reciprocal polar of a conic with regard to a circle, or with regard to any conic, is a conic.*

Taking any two tangents of the conic, their reciprocal polars are points on the reciprocal curve, and the reciprocal polar of their point of intersection is the chord joining the points.

Since only two tangents can be drawn from a point to a conic, it follows that the reciprocal curve is always intersected by a straight line in two points only.

It follows therefore that the reciprocal curve is a conic.

In reciprocating a conic with regard to a circle, the reciprocal polar is an ellipse, parabola, or hyperbola, according as the centre S of the circle is inside, upon, or outside the conic.

In the second case the axis of the parabola is parallel to the normal at the point S, and in the third case the asymptotes are perpendicular to the tangents which can be drawn from the point S to the conic.

When the auxiliary curve is a conic, centre S, the first of the preceding statements holds good.

When the point S is on the conic, the axis of the parabola is parallel to the diameter of the auxiliary conic, which is conjugate to the tangent at S.

When the point S is outside the conic, the asymptotes of the hyperbola are parallel to those diameters of the auxiliary conic which are conjugate to the straight lines through S touching the conic to be reciprocated.

The following cases will serve to illustrate the theorem of this article.

231. *The reciprocal polar of a parabola with regard to a point on the directrix is a rectangular hyperbola.*

For the two tangents from the point are at right angles to each other, and therefore the asymptotes are at right angles to each other.

232. *The reciprocal polar of an ellipse or hyperbola, with regard to its centre, is a similar curve turned through a right angle about the centre.*

If CY is the perpendicular on the tangent at P, and Q the reciprocal of the tangent, $CQ \,.\, CY$ is constant.

But $CY \,.\, CD$ is constant;

$$\therefore CQ \text{ varies as } CD,$$

and the reciprocal curve is the same as the original curve, or similar to it.

233. *The chords of a conic which subtend a right angle at a fixed point P of a conic all pass through a fixed point in the normal at P.*

Reciprocating with regard to P, the reciprocal curve is a parabola, the axis of which is parallel to the normal to the conic, and the reciprocal of the chord is the point of intersection of tangents at right angles to each other.

The locus of this point is the directrix of the parabola, and, being at right angles to the normal, it follows, on reciprocating backwards, that the chord passes through a fixed point E in the normal.

To find the position of the point E,
let C be the centre of the conic, CA, CB its semi-axes, and PNP' the double ordinate, and let the normal meet the axes in G and g.

Since CA and CB bisect the angle PCP' and its supplement,

$$C(BPAP') \text{ is an harmonic pencil;}$$

$\therefore PGEg$ is an harmonic range, so that PE is the harmonic mean between PG and Pg.

In the case of an hyperbola $EGPg$ is an harmonic range.

In the case of a parabola, E is the point of intersection of the normal with the diameter through P'.

234. *The chords of a conic which subtend a right angle at a fixed point O not on the conic all touch a conic of which that point is a focus.*

Reciprocating with regard to O, the reciprocal of the envelope of the chords is the director circle of a conic, and therefore, reciprocating backwards, it follows that the envelope of the chords is a conic of which O is a focus. This of course includes the preceding theorem as a particular case, the fact being that when O is on the conic the envelope of the chords is a conic, with a vertex and focus at E, flattened into a straight line.

235. *If the sides of a triangle are tangents to a parabola, the orthocentre of the triangle is on the directrix of the parabola.*

This theorem is at once obtained by reciprocating, with regard to the orthocentre of the triangle, the theorem, proved in Art. 143, that, if a rectangular hyperbola passes through the angular points of a triangle, it also passes through the orthocentre of the triangle.

EXAMPLES.

1. If any triangle be reciprocated with regard to its orthocentre, the reciprocal triangle will be similar and similarly situated to the original one and will have the same orthocentre.

2. If two conics have the same focus and directrix, and a focal chord be drawn, the four tangents at the points where it meets the conics intersect in the same point of the directrix.

3. An ellipse and a parabola have a common focus; prove that the ellipse either intersects the parabola in two points, and has two common tangents with it, or else does not cut it.

4. Prove that the reciprocal polar of the circumscribed circle of a triangle with regard to the inscribed circle is an ellipse, the major axis of which is equal in length to the radius of the inscribed circle.

5. Reciprocate with respect to any point S the theorem that, if two points on a circle be given, the pole of PQ with respect to that circle lies on the line bisecting PQ at right angles.

6. If two parabolas whose axes are at right angles have a common focus, prove that the part of the common tangent intercepted between the points of contact subtends a right angle at the focus.

7. The tangent at a moving point P of a conic intersects a fixed tangent in Q, and from S a straight line is drawn perpendicular to SQ and meeting in R the tangent at P; prove that the locus of R is a straight line.

8. Four parabolas having a common focus can be described touching respectively the sides of the triangles formed by four given points.

9. A triangle ABC circumscribes a parabola, focus S; through ABC lines are drawn respectively perpendicular to SA, SB, SC; shew that these lines are concurrent.

10. Prove that the distances, from the centre of a circle, of any two poles are to one another as their distances from the alternate polars.

11. Reciprocate the theorems,

(1) The opposite angles of any quadrilateral inscribed in a circle are equal to two right angles.

(2) If a line be drawn from the focus of an ellipse making a constant angle with the tangent, the locus of its intersection with the tangent is a circle.

12. The locus of the intersection of two tangents to a parabola which include a constant angle is an hyperbola, having the same focus and directrix.

13. Two ellipses having a common focus cannot intersect in more than two real points, but two hyperbolas, or an ellipse and hyperbola, may do so.

14. ABC is any triangle and P any point: four conic sections are described with a given focus touching the sides of the triangles ABC, PBC, PCA, PAB respectively; shew that they all have a common tangent.

15. TP, TQ are tangents to a parabola cutting the directrix respectively in X and Y; ESF is a straight line drawn through the focus S perpendicular to ST, cutting TP, TQ respectively in E, F; prove that the lines EY, XF are tangents to the parabola.

16. With the orthocentre of a triangle as focus, two conics are described touching a side of the triangle and having the other two sides as directrices respectively; shew that their minor axes are equal.

17. Two parabolas have a common focus S; parallel tangents are drawn to them at P and Q intersecting the common tangent in P' and Q'; prove that the angle PSQ is equal to the angle between the axes, and the angle $P'SQ'$ is supplementary.

18. ABC is a given triangle, S a given point; on BC, CA, AB respectively, points A', B', C' are taken, such that each of the angles ASA', BSB', CSC', is a right angle. Prove that A', B', C' lie in the same straight line, and that the latera recta of the four conics, which have S for a common focus, and respectively touch the three sides of the triangles ABC, $AB'C'$, $A'BC'$, $A'B'C$ are equal to one another.

19. A parabola and hyperbola have the same focus and directrix, and SPQ is a line drawn through the focus S to meet the parabola in P, and the nearer branch of the hyperbola in Q; prove that PQ varies as the rectangle contained by SP and SQ.

20. If two equal parabolas have the same focus, the tangents at points angularly equidistant from the vertices meet on the common tangent.

21. If an ellipse and a parabola have the same focus and directrix, and if tangents are drawn to the ellipse at the ends of its major axis, the diagonals of

the quadrilateral formed by the four points where these tangents cut the parabola intersect in the focus.

22. Find the reciprocals of the theorems of Arts. 215 and 217.

23. If a conic be reciprocated with regard to a point, shew that there are only two positions of the point, such that the conic may be similar and similarly situated to the reciprocal.

24. Conics are described having a common focus and equal latera recta. Also the corresponding directrices envelope a fixed confocal conic. Prove that these conics all touch two fixed conics, and that the reciprocals of the latera recta of these fixed conics are equal to the sum and difference of the latera recta of the variable conics and of the fixed confocal.

25. Given a point, a tangent, and a focus of a conic, prove that the envelope of the directrix is a conic passing through the given focus.

26. Two conics have a common focus: their corresponding directrices will intersect on their common chord, at a point whose focal distance is at right angles to that of the intersection of their common tangents.

If the conics are parabolas, the inclination of their axes will be the angle subtended by the common tangent at the common focus.

27. If the intercept on a given straight line between two variable tangents to a conic subtends a right angle at the focus of the conic, the tangents intersect on a conic.

28. The tangent at P to an hyperbola meets the directrix in Q; another point R is taken on the directrix such that QR subtends at the focus an angle equal to that between the transverse axis and an asymptote; prove that RP envelopes a parabola.

29. S is the focus of a conic; P, Q two points on it such that the angle PSQ is constant; through S, SR, ST are drawn meeting the tangents at P, Q in R, T respectively, and so that the angles PSR, QST are constant; shew that RT always touches a conic having the same focus and directrix as the original conic.

30. OA, OB are common tangents to two conics having a common focus S, CA, CB are tangents at one of their points of intersection, BD, AE tangents intersecting CA, CB, in D, E. Prove that SDE is a straight line.

31. An hyperbola, of which S is one focus, touches the sides of a triangle ABC; the lines SA, SB, SC are drawn, and also lines SD, SE, SF respectively perpendicular to the former three lines, and meeting any tangent to the curve in D, E, F; shew that the lines AD, BE, CF are concurrent.

32. If a conic inscribed in a triangle has one focus at the centre of the circumscribed circle of the triangle, its transverse axis is equal to the radius of that circle.

33. If any two diameters of an ellipse at right angles to each other meet the tangent at a fixed point P in Q and R, the other two tangents through Q and R intersect on a fixed straight line which passes through a point T on the tangent at P, such that PCT is a right angle.

CHAPTER XIII.

THE CONSTRUCTION OF A CONIC FROM GIVEN CONDITIONS.

236. It will be found that, in general, five conditions are sufficient to determine a conic, but it sometimes happens that two or more conics can be constructed which will satisfy the given conditions. We may have, as given conditions, points and tangents of the curve, the directions of axes or conjugate diameters, the position of the centre, or any characteristic or especial property of the curve.

PROP. I. *To construct a parabola, passing through three given points, and having the direction of its axis given.*

In this case the fact that the conic is a parabola is one of the conditions.

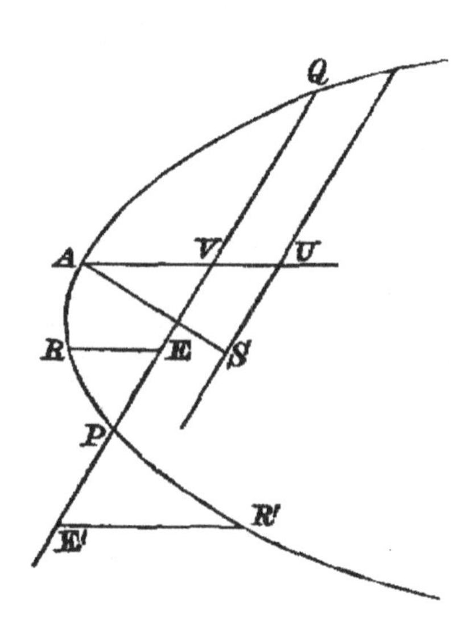

Let P, Q, R be the given points, and let RE parallel to the given direction meet PQ in E.

If E be the middle point of PQ, R is the vertex of the diameter RE; but, if not, bisecting PQ in V, draw the diameter through V and take A such that

$$AV : RE :: QV^2 : QE \, . \, EP.$$

Then A is the vertex of the diameter AV.

If the point E do not fall between P and Q, A must be taken on the side of PQ which is opposite to R.

The focus may then be found by taking AU such that

$$QV^2 = 4AV \, . \, AU,$$

and by then drawing US parallel to QV and taking AS equal to AU.

237. PROP. II. *To describe a parabola through four given points.*

First, let $ABCD$ be four points in a given parabola, and let the diameter CF meet AD in F.

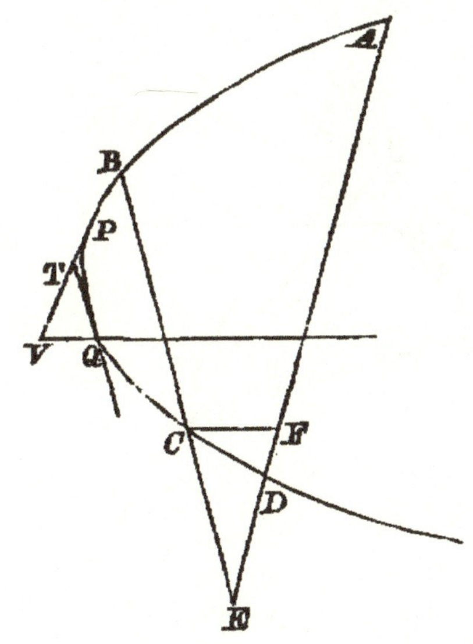

Draw the tangents PT, QT parallel to AD, BC, and the diameter QV meeting PT in V.

Then
$$\begin{aligned} ED \, . \, EA : EC \, . \, EB &:: TP^2 : TQ^2 \\ &:: TV^2 : TQ^2 \\ &:: EF^2 : EC^2. \end{aligned}$$

Hence the construction; in EA take EF such that

$$EF^2 : EC^2 :: ED \,.\, EA : EC \,.\, EB,$$

then CF is the direction of the axis, and the problem is reduced to the preceding.

If the point F be taken in AE produced, another parabola can be drawn, so that, in general, two parabolas can be drawn through four points.

238. This problem may be treated differently by help of the theorem of Art. 52, viz.;

If from a point O, outside a parabola, a tangent OM, and a chord OAB be drawn, and if the diameter ME meet the chord in E,

$$OE^2 = OA \,.\, OB.$$

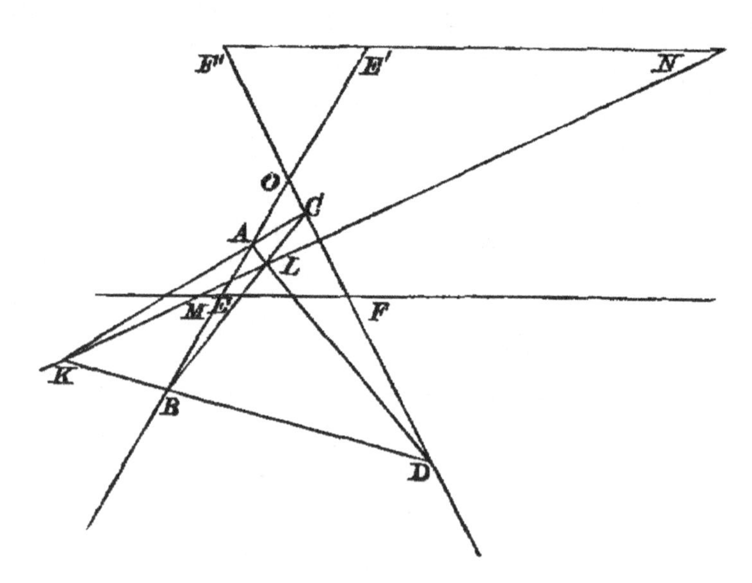

Let A, B, C, D be the given points, and let E, E', F, F', be so taken that

$$OE^2 = OE'^2 = OA \,.\, OB,$$

and

$$OF^2 = OF'^2 = OC \,.\, OD.$$

Then EF and $E'F'$ are diameters, and KL, the polar of O, will meet EF and $E'F'$ in M, N, the points of contact of tangents from O.

The second parabola is obtained by taking for diameters EF' and $E'F$.

239. PROP. III. *Any conic passing through four points has a pair of conjugate diameters parallel to the axes of the two parabolas which can be drawn through the four points.*

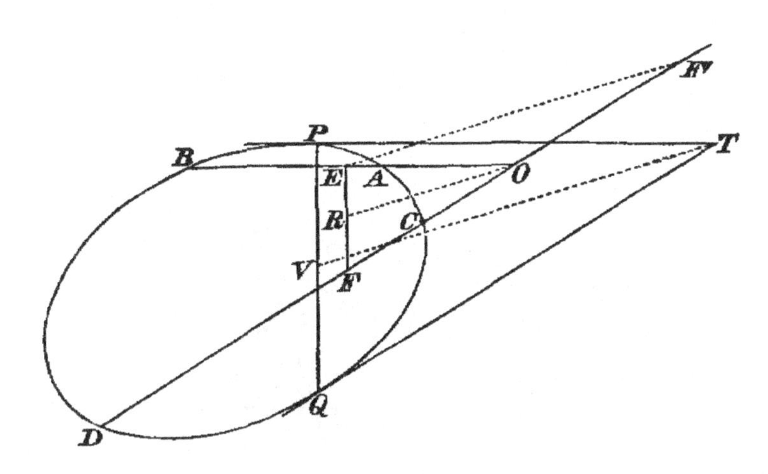

Let TP, TQ be the tangents parallel to OAB and OCD, and such that the angle PTQ is equal to AOC.

Then, if $OE^2 = OA \,.\, OB$, and $OF^2 = OC \,.\, OD$,

$$OE^2 : OF^2 :: OA \,.\, OB : OC \,.\, OD$$
$$:: TP^2 : TQ^2;$$

$\therefore EF$ is parallel to PQ.

Hence, if R and V be the middle points of EF and PQ, OR is parallel to TV;

But, taking OF' equal to OF, OR is parallel to EF',

$\therefore TV$ and PQ are parallel to EF' and EF;

i.e. the conjugate diameters parallel to TV and PQ are parallel to the axes of the two parabolas.

240. PROP. IV. *Having given a pair of conjugate diameters, PCP', DCD', it is required to construct the ellipse.*

In CP take E such that $PE \,.\, PC = CD^2$, draw PF perpendicular to CD, and take FC' equal to FC.

About CEC' describe a circle, cutting PF in G and G'; then

$$PG \,.\, PG' = PE \,.\, PC = CD^2,$$

and GCG' is a right angle; therefore CG and CG' are the directions of the axes and their lengths are given by the relations,

$$PG \,.\, PF = BC^2,$$
$$PG' \,.\, PF = AC^2.$$

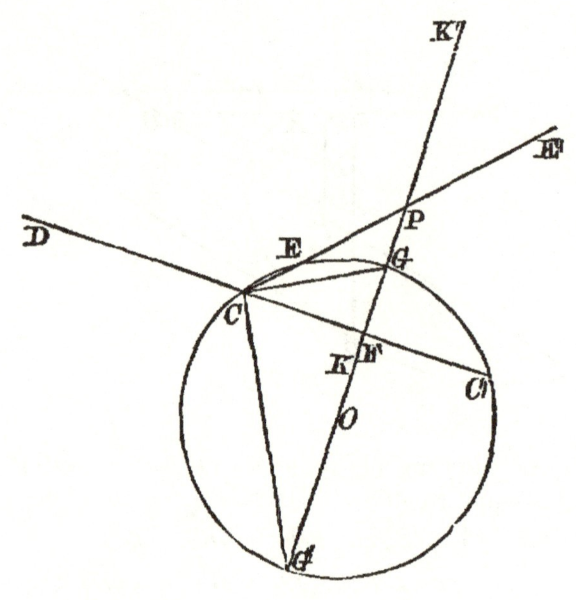

We may observe that, O being the centre of the circle,

$$\begin{aligned} AC^2 + BC^2 &= PF \,.\, PG + PF \,.\, PG' \\ &= 2 \,.\, PF \,.\, PO \\ &= 2 \,.\, PC \,.\, PN, \end{aligned}$$

if N be the middle point of CE,

$$\begin{aligned} &= PC^2 + PC \,.\, PE \\ &= CP^2 + CD^2. \end{aligned}$$

If PE' be taken equal to PE in CP produced, and the same construction be made, we shall obtain the axes of an hyperbola having CP, CD for a pair of conjugate semi-diameters.

241. This problem may be treated also as follows.

In PF, the perpendicular on CD, take

$$PK = PK' = CD;$$

then $$PK^2 = PG \,.\, PG',$$

and therefore $K'GKG'$ is an harmonic range; and GCG' being a right angle, it follows (Art. 199), that CG and CG' are the bisectors of the angles between CK and CK'.

Hence, knowing CP and CD, G and G' are determined.

242. Prop. V. *Having given the focus and three points of a conic, to find the directrix.*

Let A, B, C, S be the three points and the focus.

Produce BA to D so that

$$BD : AD :: SB : SA,$$

and CB to E, so that

$$BE : CE :: SB : SC;$$

then DE is the directrix.

The lines BA, BC may be also divided internally in the same ratio, so that four solutions are generally possible.

Conversely, if three points A, B, C and the directrix are given, let BA, BC meet the directrix in D and E; then S lies on a circle, the locus of a point, the distances of which from A and B are in the ratio of AD to DB.

S lies also on a circle, similarly constructed with regard to BCE; the intersection of these circles gives two points, either of which may be the focus.

243. Prop. VI. *Having given the centre, the directions of a pair of conjugate diameters, and two points of an ellipse, to describe the ellipse.*

If C be the centre, CA, CB the given directions, and P, Q the points, draw QMQ', PLP' parallel to CB and CA, and make $Q'M = QM$ and $P'L = PL$.

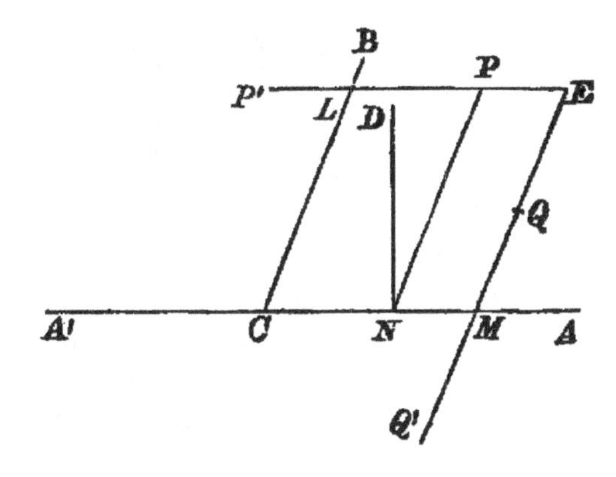

Then the ellipse will evidently pass through P' and Q', and if CA, CB be the conjugate radii, their ratio is given by the relation

$$CA^2 : CB^2 :: EP \,.\, EP' : EQ \,.\, EQ',$$

E being the point of intersection of $P'P$ and $Q'Q$.

Set up a straight line ND perpendicular to CA and such that

$$ND^2 : NP^2 :: EP \,.\, EP' : EQ \,.\, EQ',$$

and describe a circle, radius CD and centre C, cutting CA in A, and take

$$CB : CA :: NP : ND.$$

Then $$AN \,.\, NA' = ND^2,$$

and $$PN^2 : AN \,.\, NA' :: CB^2 : CA^2.$$

Hence CA, CB are determined, and the ellipse passes through P and Q.

244. Prop. VII. *To describe a conic passing through a given point and touching two given straight lines in given points.*

Let OA, OB be the given tangents, A and B the points of contact, N the middle point of AB.

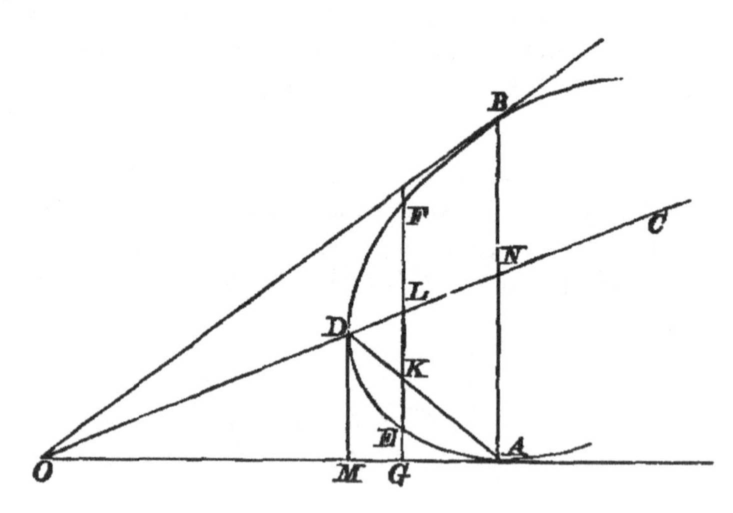

1st. Let the given point D be in ON; then, if $ND = OD$, the curve is a parabola.

But if $ND < OD$, the curve is an ellipse, and, taking C such that $OC \,.\, CN = CD^2$, the point C is the centre.

If $ND > OD$, the curve is an hyperbola, and its centre is found in the same manner.

2nd. If the given point be E, not in ON, draw GEF parallel to AB, and make FL equal to EL.

Take K such that

$$GK^2 = GE \,.\, GF;$$

then AK produced will meet ON in D, and the problem is reduced to the first case.

To justify this construction, observe that, if DM be the tangent at D,

$$GE \,.\, GF : GA^2 :: DM^2 : MA^2$$
$$:: GK^2 : GA^2,$$

so that

$$GE \,.\, GF = GK^2.$$

245. PROP. VIII. *To draw a conic through five given points.*

Let A, B, C, D, E be the five points, and F the intersection of DE, AB.

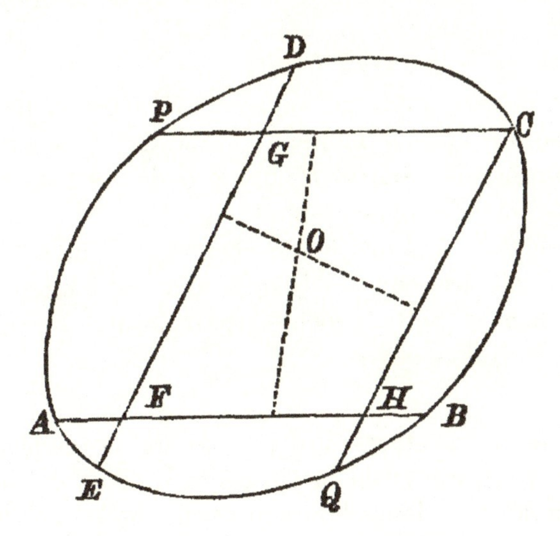

Draw CG, CH, parallel respectively to AB and ED, and meeting ED, AB in G and H.

If F and G fall between D and E, and F and H between A and B, take GP in CG produced and HQ in CH produced, such that

$$CG \,.\, GP : DG \,.\, GE :: AF \,.\, FB : DF \,.\, FE,$$

and

$$CH \,.\, HQ : AH \,.\, HB :: DF \,.\, FE : AF \,.\, FB;$$

Then (Arts. 92 and 134) P and Q are points in the conic.

Also PC, AB being parallel chords, the line joining their middle points is a diameter, and another diameter is obtained from CQ and DE.

If these diameters are parallel, the conic is a parabola, and we fall upon the case of Prop. II.; but if they intersect in a point O, this point is the centre of the conic, and, having the centre, the direction of a diameter, and two ordinates of that diameter, we fall upon the case of Prop. VI.

The figure is drawn for the case in which the pentagon $AEBCD$ is not re-entering, in which case the conic may be an ellipse, a parabola, or an hyperbola.

If any one point fall within the quadrilateral formed by the other four, the curve is an hyperbola.

In all cases the points P, Q must be taken in accordance with the following rule.

The points C, P, or C, Q must be on the same or different sides of the points G, or H, according as the points D, E, or B, A are on the same or different sides of the points G or H.

Thus, if the point E be between D and F, and if G be between D and E, and H between A and B, the points P and C will be on the same side of G, and C, Q on the same side of H, but if H do not fall between A and B, C and Q will be on opposite sides of H.

Remembering that if a straight line meet only one branch of an hyperbola, any parallel line will meet only one branch, and that if it meet both branches, any parallel will meet both branches, the rule may be established by an examination of the different cases.

246. The above construction depends only on the elementary properties of Conics, which are given in Chapters I., II., III., and IV. For some further constructions we shall adopt another method depending on harmonic properties.

PROP. IX. *Having given two pairs of lines OA, OA', and OB, OB', to find a pair of lines OC, OC', which shall make with each of the given pairs an harmonic pencil.*

This is at once effected by help of Art. 203.

For, if any transversal cut the lines in the points c, a, b, c', b', a', the points c, c' are the foci of the involution, in which a, a' are conjugate, and also b, b', the centre of the involution being the middle point of cc'.

247. PROP. X. *If two points and two tangents of a conic be given, the chord of contact intersects the given chord in one of two fixed points*.*

Let OP, OQ be the given tangents, A and B the given points, and C the intersection of AB and the chord of contact.

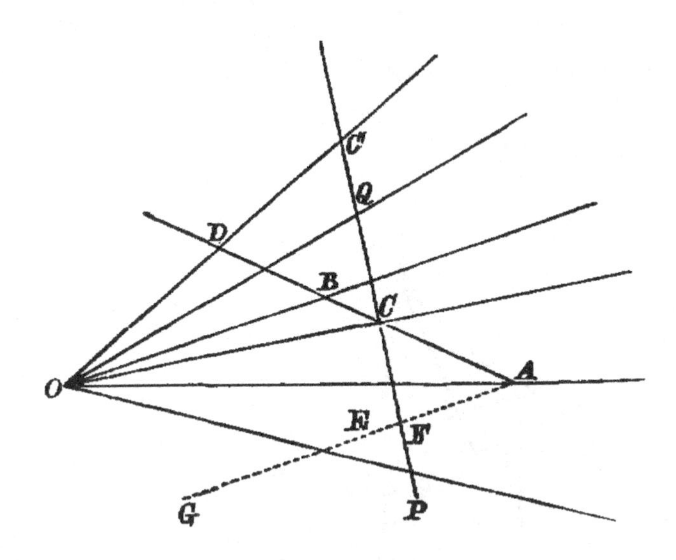

Let OC' be the polar of C, and let AB meet OC' in D.

Then C is on the polar of D, and therefore $DBCA$ is an harmonic range.

Also, C being on the polar of C', $C'QCP$ is an harmonic range.

Hence if two lines OC, OC' be found, which are harmonic with OA, OB, and also with OP, OQ, these lines intersect AB in two points C and D, through one of which the chord of contact must pass.

Or thus, if the tangents meet AB in a and b, find the foci C and D of the involution AB, ab; the chord of contact passes through one of these points.

248. PROP. XI. *Having given three points and two tangents, to find the chord of contact.*

In the preceding figure let OP, OQ be the tangents, and A, B, E the points.

Find OC, OC' harmonic with OA, OB, and OP, OQ; also find OF, OG harmonic with OA, OE and OP, OQ.

---

*I am indebted to Mr Worthington for much valuable assistance in this chapter, and especially for the constructions of Articles 247, 249, 250, and 253.

Then any one of the four lines joining C or D to F or G is a chord of contact, and the chord of contact and points of contact being known, the case reduces to that of Art. 244.

Hence four such conics can in general be described.

249. Prop. XII. *To describe a conic, passing through two given points, and touching three given straight lines.*

Let AB, the line joining the given points, meet the given tangents QR, RP, PQ, in N, M, L.

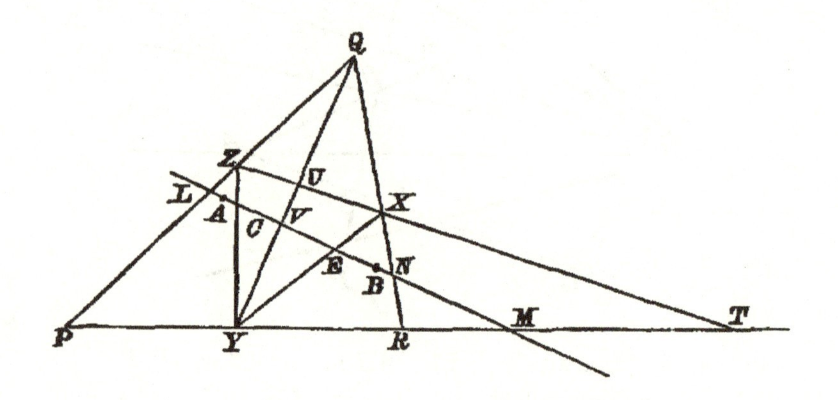

Find the foci C, D of the involution A, B and L, M;

Then YZ, the polar of P, passes through C or D, Art. 247.

Also find the foci, E, F, of the involution A, B, and M, N; then XY, the polar of R, passes through F or E.

Let ZX meet PR in T; then T is on the polar of Q, and QY is the polar of T.

Hence $\qquad TXUZ$ is harmonic;

therefore $\qquad MEVC$ is harmonic.

This determines V, and, joining QV, we obtain the point of contact Y.

Then, joining YC and YE, Z and X are obtained, and X, Y, Z being points of contact, we have five points, and can describe the conic by the construction of Art. 245, or by that of Art. 252.

Since either C or D may be taken with E or F, there are in general four solutions of the problem.

250. PROP. XIII. *To describe a conic, having given four points and one tangent.*

Let A, B, C, D be the given points, and complete the quadrilateral.

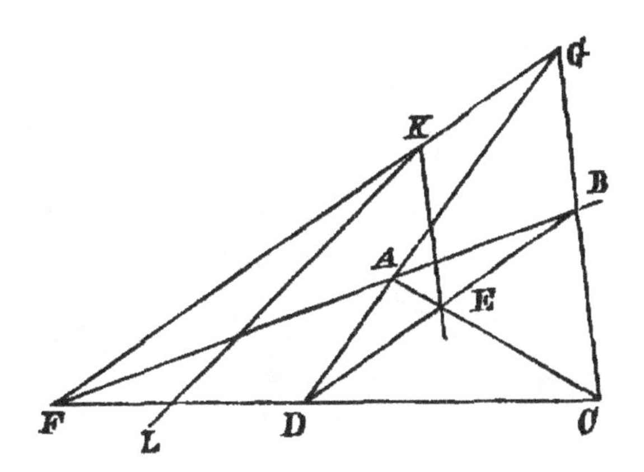

Then E is the pole of FG, and if the given tangent KL meet FG in K, E is on the polar of K; therefore the other tangent through K forms an harmonic pencil with KF, KL, KE.

Hence two tangents being known, and a point E in the chord of contact, if we find two points P, P' in A, B, such that KP, KP' are harmonic with KA, KB, and also with KL, KL', we shall have two chords of contact EP, EP', and therefore two points of contact for KL and also for KL'.

Hence two conics can be described.

We observe that if two conics pass through four points, their common tangents meet on one of the sides of the self-conjugate triangle EFG.

251. PROP. XIV. *Given four tangents and one point, to construct the conic.*

Let $ABCD$ be the given circumscribing quadrilateral, and E the given point. Completing the figure, draw LEF through E and F, and complete the harmonic range $LEFE'$; then, since F is the pole of HG (Art. 217), E' is a point in the conic.

Also, since K is the pole of FA (Art. 217), the chord of contact of the tangents AB, AD, passes through K.

Hence the construction is the same as that of Art. 250, and there are two solutions of the problem.

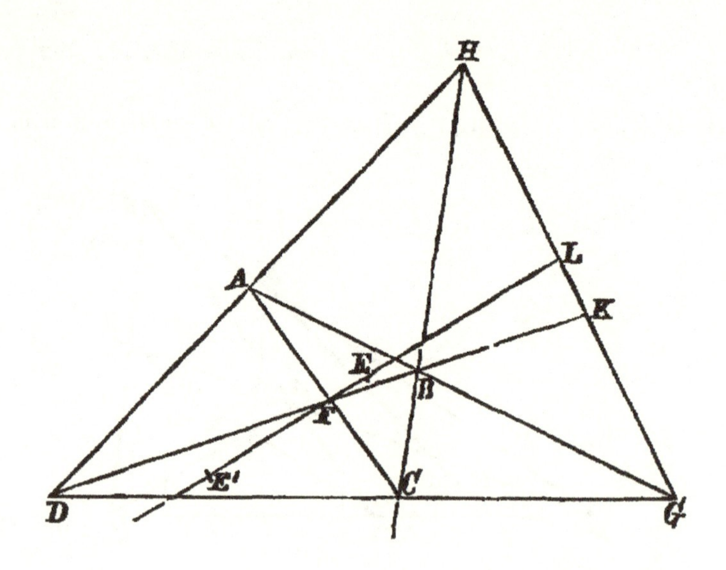

252. PROP. XV. *Given five points, to construct the conic.*

Let A, B, C, D, E be the five points, and complete the quadrilateral $ABCD$.

Then H is the pole of FG, and FG passes through the points of contact P, Q of the tangents from H.

Join HE, cutting FG in K, and complete the harmonic range $HEKE'$; then E' is a point in the conic.

Also AE, BE' will intersect FG in the same point F', and $E'A$, EB will also intersect FG in the same point G'.

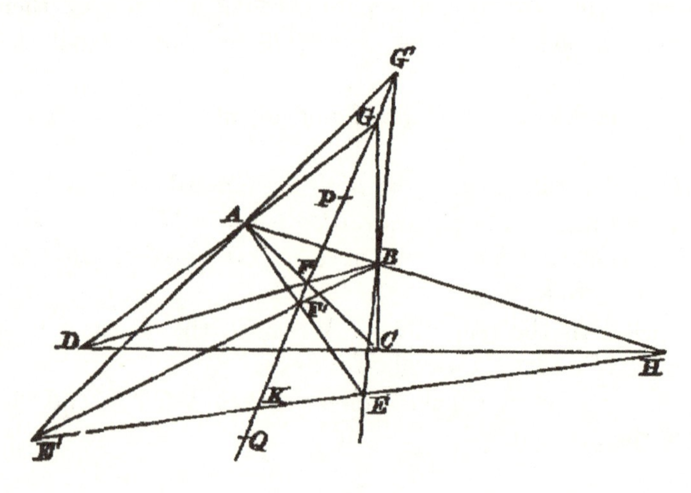

But $GPFQ$ and $G'PF'Q$ are both harmonic ranges, therefore P and Q are the foci of an involution of which F, G and F', G' are pairs of conjugate points.

Hence, finding these foci, P and Q, the tangents HP, HQ are known, and the case is reduced to that of Prop. VII.

Hence only one conic can be drawn through five points.

253. PROP. XVI. *Given five tangents, to find the points of contact.*

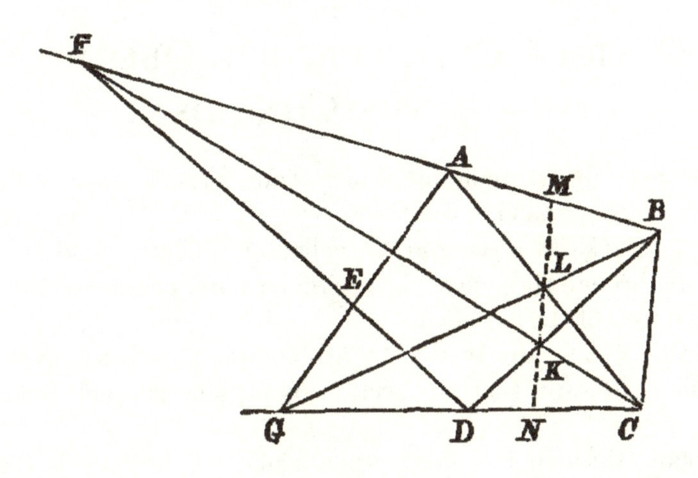

Let $ABCDE$ be the circumscribing pentagon. Considering the quadrilateral $FBCD$, join FC, BD, meeting in K.

Then (Art. 217) K is the pole of the line joining the intersections of FB, CD, and of FD, BC; that is, the chords of contact of BF, CD, and of BC, FD meet in K.

Similarly if BG, AC meet in L, the chords of contact of AB, CG, and of BC, AG meet in L.

Hence KL is the chord of contact of AB, CD, and therefore determines M, N the points of contact.

Hence it will be seen that only one conic can be drawn touching five lines.

CHAPTER XIV.

The Oblique Cylinder, the Oblique Cone, and the Conoids.

254. Def. If a straight line, which is not perpendicular to the plane of a given circle, move parallel to itself, and always pass through the circumference of the circle, the surface generated is called an oblique cylinder.

The line through the centre of the circular base, parallel to the generating lines, is the axis of the cylinder.

It is evident that any section by a plane parallel to the axis consists of two parallel lines, and that any section by a plane parallel to the base is a circle.

The plane through the axis perpendicular to the base is the principal section.

The section of the cylinder by a plane perpendicular to the principal section, and inclined to the axis at the same angle as the base, is called a subcontrary section.

255. Prop. I. *The subcontrary section of an oblique cylinder is a circle.*

The plane of the paper being the principal plane and APB the circular base, a subcontrary section is DPE, the angles BAE, DEA being equal.

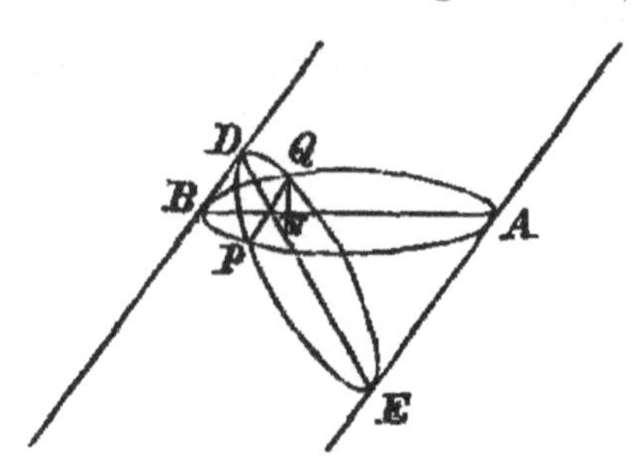

Let PQ be the line of intersection of the two sections; then

$$PN \,.\, NQ \text{ or } PN^2 = BN \,.\, NA.$$

But $\quad NB = ND$, and $NA = NE$;

$$\therefore PN \,.\, NQ = DN \,.\, NE,$$

and DPE is a circle.

256. PROP. II. *The section of an oblique cylinder by a plane which is not parallel to the base or to a subcontrary section is an ellipse.*

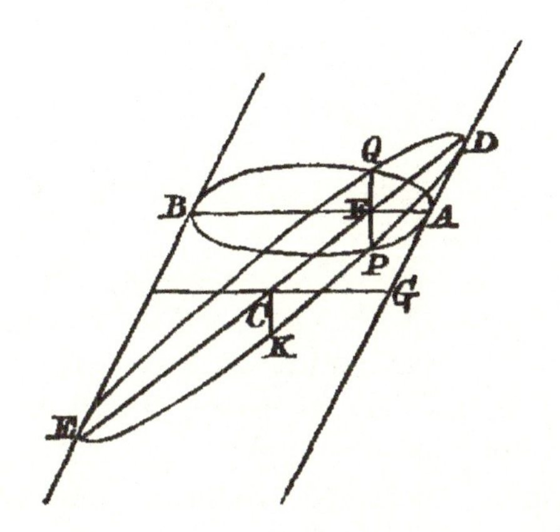

Let the plane of the section, DPE, meet any circular section in the line PQ, and let AB be that diameter of the circular section which is perpendicular to PQ, and bisect PQ in the point F.

Let the plane through the axis and the line AB cut the section DPE in the line DPE.

Then $$PF^2 = AF \,.\, FB.$$

But if DE be bisected in C, and GKC be the circular section through C parallel to APB,

$$AF : FD :: CG : CD,$$

and $$FB : FE :: CG : CD;$$

$$\therefore AF \,.\, FB : DF \,.\, FE :: CG2 : CD2;$$

hence, observing that $CG = CK$,

$$PF^2 : DF \,.\, FE :: CK^2 : CD^2.$$

But, if a series of parallel circular sections be drawn, PQ is always parallel to itself and bisected by DE;

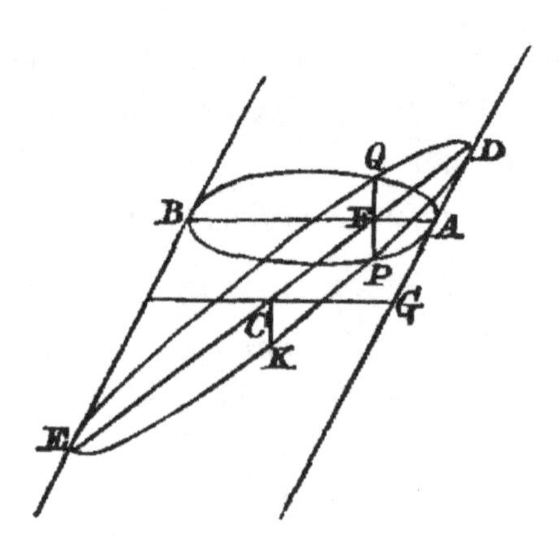

Therefore the curve DPE is an ellipse, of which CD, CK are conjugate semi-diameters.

257. DEF. If a straight line pass always through a fixed point and the circumference of a fixed circle, and if the fixed point be not in the straight line through the centre of the circle at right angles to its plane, the surface generated is called an oblique cone.

The plane containing the vertex and the centre of the base, and also perpendicular to the base, is called the principal section.

The section made by a plane not parallel to the base, but perpendicular to the principal section, and inclined to the generating lines in that section at the same angle as the base, is called a subcontrary section.

258. PROP. III. *The subcontrary section of an oblique cone is a circle.*

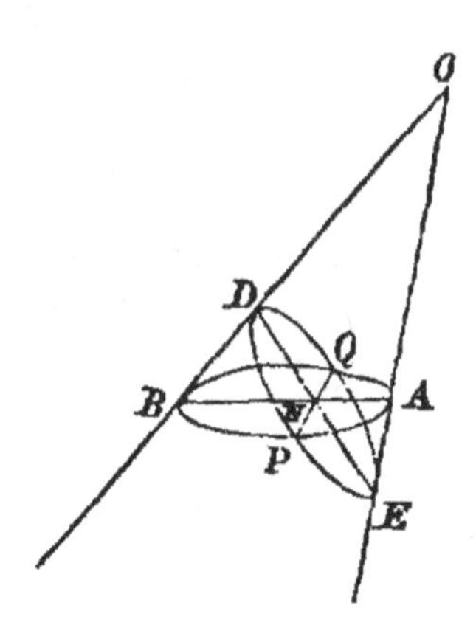

The plane of the paper being the principal section, let APB be parallel to the base and DPE a subcontrary section, so that the angle

$$ODE = OAB,$$

and $$OED = OBA.$$

The angles DBA, DEA being equal to each other, a circle can be drawn through $BDAE$.

Hence, if PNQ be the line of intersection of the two planes APB and EPD,

$$DN \,.\, NE = BN \,.\, NA,$$
$$= PN \,.\, NQ;$$

therefore DPE is a circle.

And all sections by planes parallel to DPE are circles.

Planes parallel to the base, or to a subcontrary section, are called also *Cyclic Planes.*

259. PROP. IV. *The section of a cone by a plane not parallel to a cyclic plane is an Ellipse, Parabola, or Hyperbola.*

(1) Let the section, DPE, meet all the generating lines on one side of the vertex.

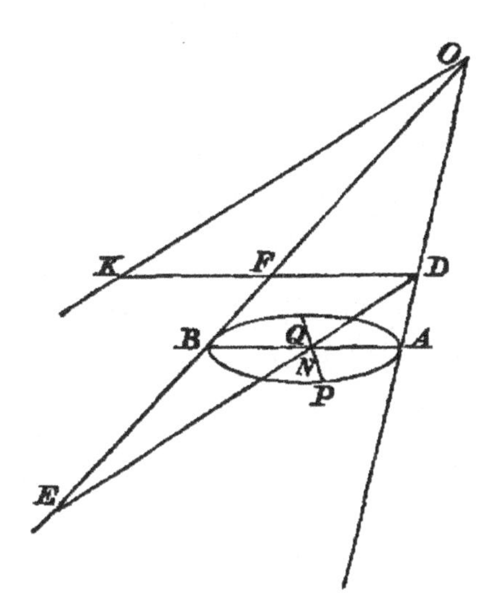

Let any circular section cut DPE in PQ, and take AB the diameter of the circle which bisects PQ.

The plane OAB will cut the plane of the section in a line DNE.

Draw OK parallel to DE and meeting in K the plane of the circular section through D parallel to APB, and join DK, meeting OE in F.

Then $$AN : ND :: KD : OK,$$

and $$BN : NE :: KF : OK;$$

therefore $$AN . NB : DN . NE :: KD . KF : OK^2,$$

or $$PN^2 : DN . NE :: KD . KF : OK^2.$$

But if a series of circular sections be drawn the lines PQ will always be parallel, and bisected by DE;

Therefore the curve DPE is an ellipse, having DE for a diameter, and the conjugate diameter parallel to PQ, and the squares on these diameters are in the ratio of $KD . KF$ to OK^2.

(2) Let the section be parallel to a tangent plane of the cone.

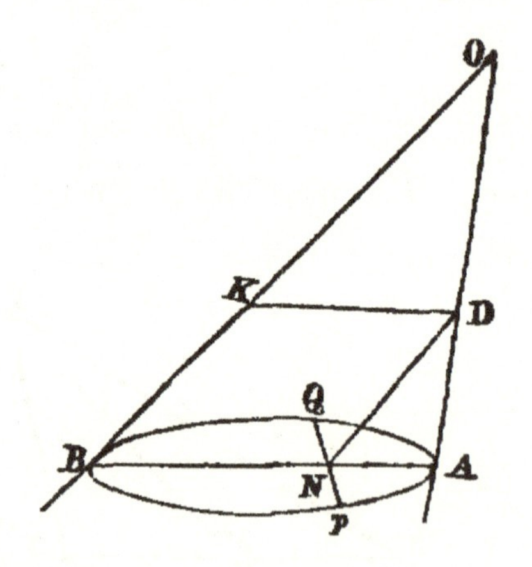

If OB be the generating line along which the tangent plane touches the cone, and BT the tangent line at B to a circular section through B, the line of intersection PQ will be parallel to BT, and therefore perpendicular to the diameter BA through B.

Let the plane BOA cut the plane of the section in DN.

Then, drawing DK parallel to AB,

$$BN = KD,$$

and $$AN : ND :: KD : OK;$$

therefore $$AN . NB : ND . KD :: KD : OK,$$

or $$PN^2 : ND . KD :: KD : OK,$$

and KD, OK being constant, the curve is a parabola having the tangent at D parallel to PQ.

If the plane of the section meet both branches of the cone, make the same construction as before, and we shall obtain, in the same manner as for the ellipse,

$$PN^2 : DN \,.\, NE :: DK \,.\, KF : OK^2,$$

OK being parallel to DE.

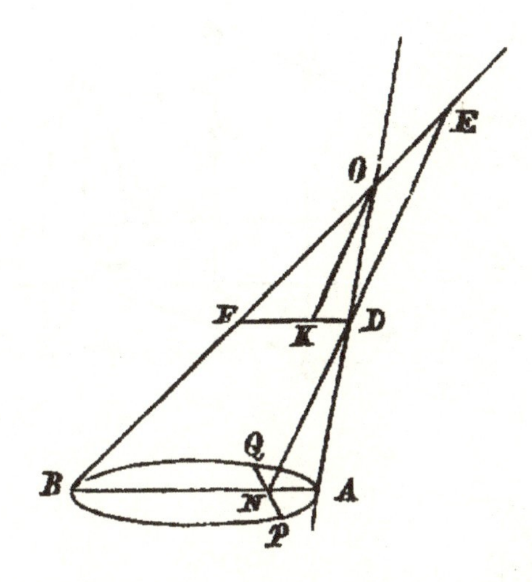

Therefore, since the point N is not between the points D and E, the curve DP is an hyperbola.

Conoids.

260. DEF. *If a conic revolve about one of its principal axes, the surface generated is called a conoid.*

If the conic be a circle, the conoid is a sphere.

If the conic be an ellipse, the conoid is an oblate or a prolate spheroid according as the revolution takes place about the conjugate or the transverse axis.

If it be an hyperbola the surface is an hyperboloid of one or two sheets, according as the revolution takes place about the conjugate or transverse axis, and the surface generated by the asymptotes is called the asymptotic cone.

If the conic consist of two intersecting straight lines, the limiting form of an hyperbola, the revolution will be about one of the lines bisecting the angles between them, and the conoid will then be a right circular cone.

261. PROP. V. *A section of a paraboloid by a plane parallel to the axis is a parabola equal to the generating parabola, and any other section not perpendicular to the axis is an ellipse.*

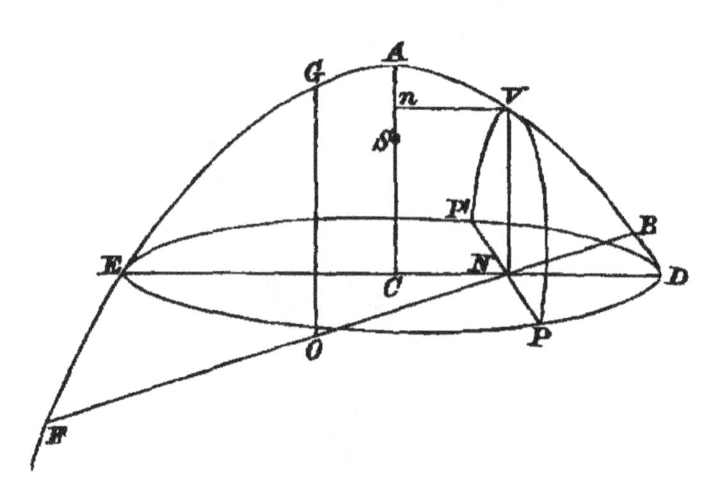

Let PVN be a section parallel to the axis, and take the plane of the paper perpendicular to the section and cutting it in VN.

Take any circular section DPE, cutting the section PVN in PNP'.

Then PN is perpendicular to DE,

and
$$\begin{aligned} PN^2 &= DN \,.\, NE \\ &= DC^2 - NC^2 \\ &= 4AS \,.\, AC - 4AS \,.\, An \\ &= 4AS \,.\, VN; \end{aligned}$$

therefore the curve VP is a parabola equal to EAD.

Again, let BPF be a section not parallel or perpendicular to the axis, but perpendicular to the plane of the paper;

Then, $BN \,.\, NF = 4SG \,.\, VN$, OG being the diameter bisecting BF (Art. 51);

therefore
$$PN^2 : BN \,.\, NF :: AS : SG,$$

and the curve BPN is an ellipse.

Moreover if the plane BF move parallel to itself, SG is unaltered, and the *sections by parallel planes are similar ellipses.*

In exactly the same manner, it may be shewn that the oblique sections of spheroids are ellipses, and those of hyperboloids either ellipses or hyperbolas.

262. Prop. VI. *The sections of an hyperboloid and its asymptotic cone by a plane are similar curves.*

Taking the case of an hyperboloid of two sheets, let DPF, $dP'f$, be the sections of the hyperboloid and cone, $P'PN$ the line in which their plane is cut by a circular section GPK or $gP'k$.

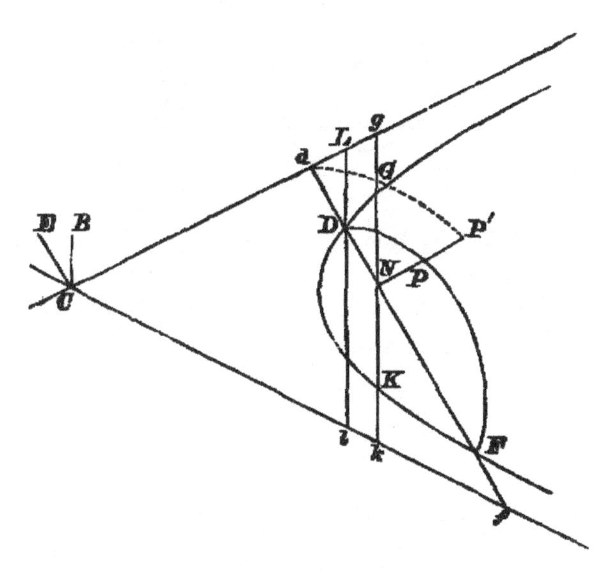

Through D draw LDl perpendicular to the axis; then, since

$$PN^2 = GN \,.\, NK, \text{ and } P'N^2 = gN \,.\, Nk,$$

$$P'N^2 : dN \,.\, Nf :: gN \,.\, Nk : dN \,.\, Nf,$$

$$:: LD \,.\, lD : Dd \,.\, Df,$$

$$:: BC^2 : CE^2$$

if CE be the semi-diameter parallel to DF;

and $$PN^2 : DN \,.\, NF :: GN \,.\, NK : DN \,.\, NF$$

$$:: BC^2 : CE^2 (\text{Art. } 134);$$

therefore the curves DPF, $dP'f$ have their axes in the same ratio, and are similar ellipses.

In the same manner the theorem can be established if the sections be hyperbolic, or if the hyperboloid be of one sheet.

263. PROP. VII. *If an hyperboloid of one sheet be cut by a tangent plane of the asymptotic cone, the section will consist of two parallel straight lines.*

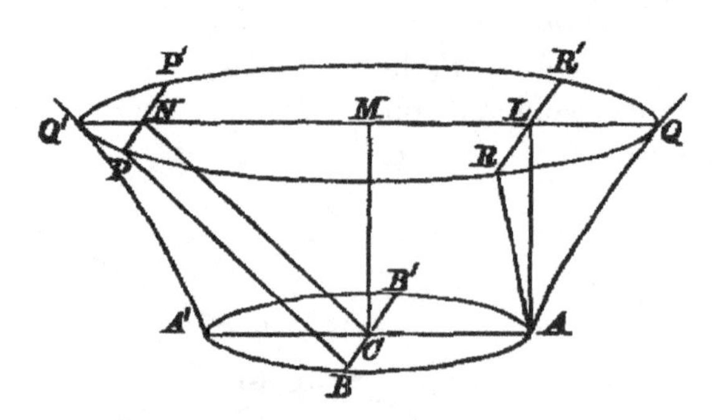

Let AQ, $A'Q'$ be a section through the axis, CN the generating line, in the plane CAQ, along which the tangent plane touches the cone; and PNP' the section with this tangent plane of a circular section QPQ'.

Then $$PN^2 = QN \,.\, NQ'$$
$$= AC^2 (\text{Art. } 106) = BC^2,$$

therefore, if BCB' be the diameter, perpendicular to the plane CAQ, of the principal circular section,

$$PN = BC \text{ and } P'N = B'C;$$

therefore PB and $P'B'$ are each parallel to CN; that is, the section consists of two parallel straight lines.

264. PROP. VIII. *The section of an hyperboloid of one sheet by a plane parallel to its axis, and touching the central circular section, consists of two straight lines.*

Let the plane pass through A, and be perpendicular to the radius CA of the central section (fig. Art. 263).

The plane will cut the circular section QPQ' in a line RLR', and

$$RL^2 = QL \,.\, LQ' = QM^2 - AC^2,$$

if M be the middle point of QQ'.

But $$QM^2 - AC^2 : CM^2 :: AC^2 : BC^2;$$

therefore $$RL : AL :: AC : BC;$$

hence it follows that AR is a fixed line; and similarly AR' is also a fixed line.

It will be seen that these lines are parallel to the section of the cone by the plane through the axis perpendicular to CA.

265. PROP. IX. *If a conoid be cut by a plane, and if spheres be inscribed in the conoid touching the plane, the points of contact of the spheres with the plane will be the foci of the section, and the lines of intersection of the planes of contact with the plane of section will be the directrices.*

In order to establish this statement, we shall first demonstrate the following theorem;

If a circle touch a conic in two points, the tangent from any point of the conic to the circle bears a constant ratio to its distance from the chord of contact.

Take the case of an ellipse, the chord of contact being perpendicular to the transverse axis.

If EME' be this chord, the normal EG is the radius of the circle, and if PT be a tangent from a point P of the ellipse,

$$PT^2 = PG^2 - GE^2$$
$$= PN^2 + NG^2 - EM^2 - MG^2.$$

But $$EM^2 - PN^2 : CN^2 - CM^2 :: BC^2 : AC^2,$$

and $$CN^2 - CM^2 = MN(CM + CN).$$

Let the normal at P meet the axis in G';

then $$NG' : CN :: BC^2 : AC^2,$$

and $$MG : CM :: BC^2 : AC^2;$$

therefore $$NG' + MG : CN + CM :: BC^2 : AC^2.$$

Hence $$EM^2 - PN^2 = MN(NG' + MG).$$

Also $$NG^2 - MG^2 = MN(NG + MG);$$

therefore $$PT^2 = MN(NG + MG) - MN(NG' + MG)$$
$$= MN \,.\, GG'.$$

But $$CG : CM :: SC^2 : AC^2,$$

and $$CG' : CN :: SC^2 : AC^2;$$

therefore $$GG' : MN :: SC^2 : AC^2.$$

Hence $$PT^2 : PL^2 :: SC^2 : AC^2,$$

PL being equal to MN.

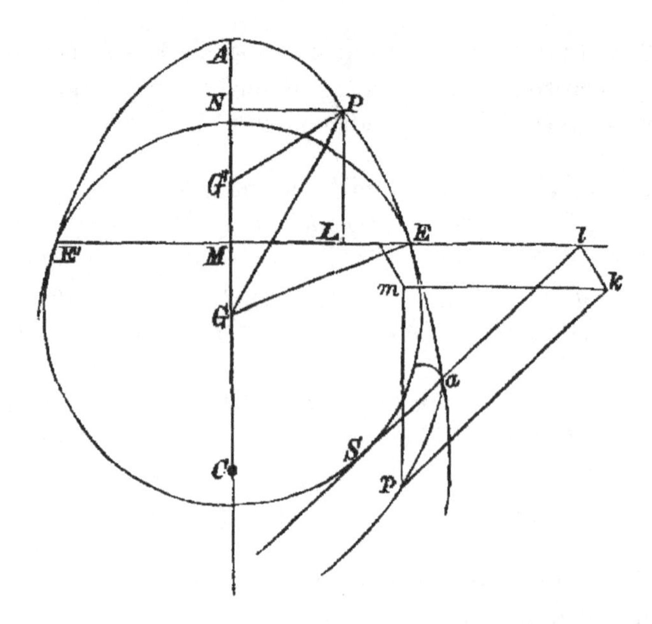

This being established let the figure revolve round the axis AC, and let a plane section ap of the conoid, perpendicular to the plane of the paper, touch the sphere at S and cut the plane of contact EE' in lk.

From a point p of the section let fall the perpendicular pm on the plane EE', draw mk perpendicular to lk, and join pk.

Then $pm : pk$ is a constant ratio.

Also taking the meridian section through p, pS is equal to the tangent from p to the circular section of the sphere, and is therefore in a constant ratio to pm;

Hence Sp is to pk in a constant ratio,
and therefore S is the focus and kl the directrix of the section ap.

266. If the curve be a parabola focus S', the proof is as follows:

$$
\begin{aligned}
PT^2 &= PG^2 - EG^2 \\
&= PN^2 + NG^2 - EM^2 - MG^2 \\
&= MN(NG + MG) - 4AS' \,.\, MN \\
&= MN(NG + MG) - 2MG \,.\, MN \\
&= MN^2.
\end{aligned}
$$

It will be found that the theorem is also true for an hyperboloid of two sheets, and for an hyperboloid of one sheet, but that in the latter case the constant ratio of PT to PL is not that of SC to AC.

267. The geometrical enunciation of the theorem also requires modification in several cases. To illustrate the difficulty, take the paraboloid, and observe that if the normal at E cuts the axis in G, and if O be the centre of curvature at A,

$$AG > AO,$$

and the radius of the circle is never less than AO.

This shews that a circle the radius of which is less than AO cannot be drawn so as to touch the conic in two points.

We may mention one exceptional case in which the theorem takes a simple form.

In general

$$EG^2 = EM^2 + MG^2 = 4AS'(AM + AS')$$
$$= 4AS' \,.\, S'G.$$

Taking the point g between S' and O, describe a circle centre g and such that the square on its radius $= 4AS' \,.\, S'g$.

Also take a point F in the axis produced such that

$$AF = Og;$$

it will then be found that the tangent from P to the circle will be equal to NF.

When g coincides with S', the circle becomes a point,

and $$AF = AS';$$

we thus fall back on the fundamental definition of a parabola.

It will be found that if the plane section of the conoid pass through S', the point S' is a focus of the section.

CHAPTER XV.

CONICAL PROJECTION.

268. If from any fixed point straight lines are drawn to all the points of a figure, the section by any plane of the lines thus drawn is the conical projection of the figure upon that plane.

The fixed point is called the vertex of projection, and the plane is called the plane of projection.

Taking the eye as the vertex of projection, the conical projection of any figure upon a plane is a perspective drawing of that figure as seen by the eye.

A straight line is projected into a straight line, for the plane through the vertex and the straight line intersects the plane of projection in a straight line.

A tangent to a curve is projected into a tangent to the projection of the curve, for two consecutive points of a curve project into two consecutive points.

Hence it follows that *a pole and polar project into a pole and polar.*

Again, *the degree of a curve is unaltered by projection*, for any number of collinear points project into the same number of collinear points.

In particular, the projection of a conic on any plane is a conic.

269. *Any straight line in a figure can be projected to an infinite distance.*

This is effected by taking the plane of projection parallel to the plane through the vertex of projection and the straight line.

270. *A system of concurrent straight lines in a plane can be projected into a system of parallel straight lines, and a system of parallel straight lines can be projected into a system of concurrent straight lines.*

The first of these is effected by taking for plane of projection any plane parallel to the straight line joining the vertex of projection and the point of concurrence.

The second is effected by taking for plane of projection any plane not parallel to the direction of the parallel straight lines.

271. *Any angle in a plane can be projected, on any other plane, into any other angle.*

Let ACB be the angle to be projected, and let DEF be the plane upon which it is to be projected.

Take any plane parallel to DEF, intersecting in A and B the lines forming the angle ACB, and take any point O in the plane.

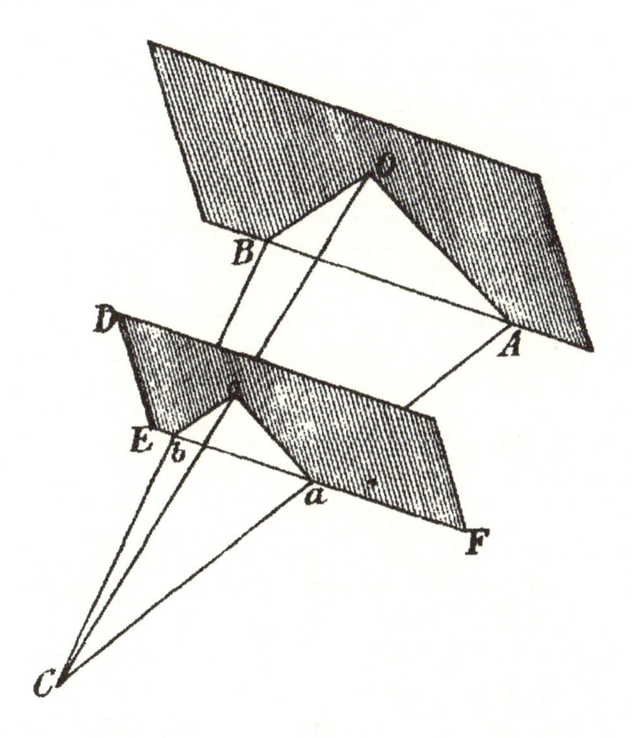

Then, if CA, CB, CO, meet the plane of projection in a, b, c, the angle acb is the projection of the angle ACB from the vertex O upon the plane DEF.

Now OA, OB are parallel to ca, cb; therefore the angle acb is equal to the angle AOB.

If then we describe on AB an arc of a circle containing an angle equal to any given angle, and take any point O on the arc as vertex of projection, the angle ACB will be projected into the given angle.

It will be seen that the arc of a circle may be described on the other side of the plane CAB, so that the locus of O on the plane OAB consists of two equal arcs on the same base.

If the plane of projection be assigned, it follows, since the plane OAB may be taken at any distance from C, that the locus of O consists of portions of two oblique cones having their common vertex at C.

If the plane of projection be not assigned, but if the line AB be assigned, the locus of O will be the surface generated by the revolution, about AB, of the arc of the circle.

If the angle ACB is to be projected into a right angle, the locus of O will be the sphere described upon AB as diameter.

If the assigned plane, DEF, be parallel to CA, the locus of O on the plane OBA will be the straight line BO making with BA the angle OBA equal to the supplement of the angle into which ACB is to be projected.

In the particular case in which this angle is a right angle the locus of O will be the straight line BO perpendicular to BA.

If it be required to project two given angles in a plane into two other given angles in any other plane, we can construct two arcs of circles in a plane parallel to this other plane, and, if these arcs intersect, the position of O is determined.

272. *To project a given quadrilateral into a square.*

Let $ABCD$ the quadrilateral, and let AC, BD intersect in E, AD, BC in F, and BA, CD in G.

Then if O is the vertex of projection, taken anywhere, the quadrilateral will be projected into a parallelogram on any plane parallel to OFG.

If O be taken on the sphere of which FG is diameter, the projection on any plane parallel to OFG will be a rectangle, for the angles subtended by FG at A, B, C, D project into right angles.

If AC and BD meet FG in L and M, and if O be taken on the circle which is the intersection of the spheres on FG and LM as diameters, the angle LEM will be projected into a right angle, so that the projection of $ABCD$ will be a rectangle, the diagonals of which are at right angles, and therefore will be a square.

273. *The projection of an harmonic range is an harmonic range.*

This is proved in Art. 198.

The projection of a circle is a conic.

This is proved in Art. 259.

As an illustration it is easily shown for a circle that, if a diameter pP passes through an external point T and intersects in V the polar of T, $pVPT$ is an harmonic range.

By projection we at once obtain the theorems of Art. 78 and of Art. 117.

274. *To project a conic into a circle, so that the projection of a given point inside the conic shall be the centre of the projection.*

Let E be the given point, AEB the chord bisected at E and PEp the diameter passing through E.

Then, if we project the polar of E to an infinite distance, and the angles AEP, APB into right angles, the projection of the conic will be a circle, the centre of which is the projection of the point E.

For the centre is the pole of a line at an infinite distance, and, the projection of AEP being a right angle, the projections of AB and Pp are the principal axes of the projection.

Also, the projection of APB being a right angle, it follows that the projection of the conic is a circle.

Another method will be to take points C, C', D, D' on the polar of E, such that CED, $C'ED'$ are self-conjugate triangles, and then to project CD to an infinite distance and the angles CED, $C'ED'$ into right angles.

The projection will be a conic, having the projection of E for its centre, and also having two pairs of conjugate diameters at right angles to each other; that is, it will be a circle.

In a subsequent article this question will be treated in a different manner.

If the point E is outside the conic, we can project the conic into a rectangular hyperbola, of which the projection of E is the centre.

For, if PQ is the chord of contact of tangents from E, all we have to do is to project PQ to an infinite distance, and PEQ into a right angle.

We can also project the conic into an hyperbola of any given eccentricity.

For, if the eccentricity is given, the angle between the asymptotes is given, and we can project PQ to an infinite distance and PEQ into the given angle.

275. *To project a conic on a given plane so that the projection of a point S inside the conic shall be a focus of the projection.*

Let the tangent at any point P and any straight line through S meet the polar of S in F and X.

Then, if we project the angles SXF, FSP into right angles, the projections of S and FX are the focus and directrix of the projection.

If at the same time we project to an infinite distance the polar of any point E on XS, the projection of E will be the centre of the projection of the conic.

276. *If two conics in different planes have two points in common, two cones of the second order can be drawn passing through them, or, in other words, each can be projected into the other.*

Let AB be the common chord, F and D its poles with regard to the conics.

Take any point E in AB, and let the plane FED meet the conics in the points P, p, Q, q, and let pq intersect DF in O.

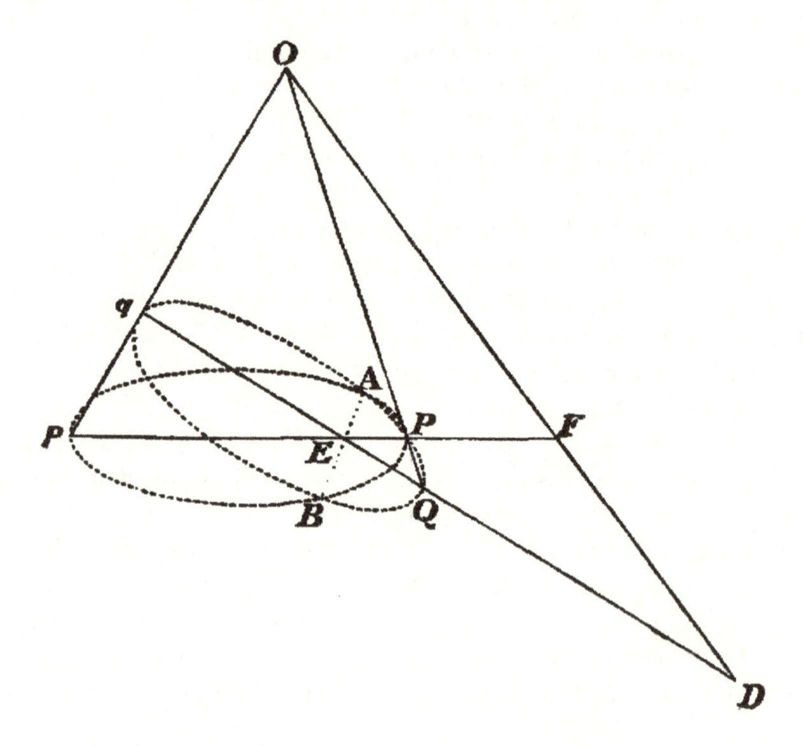

If from O the conic $BPAp$ be projected on to the plane of the other conic, the projection will be a conic touching the conic $BQAq$ at A and B, so that it will have four points in common with $BQAq$, and will also have the point q in common with $BQAq$.

Now it is proved in Art. 252, that only one conic can be drawn through five points.

Hence the projection, having five points in common with $BQAq$, coincides with it entirely.

It will be observed that OPQ is a straight line, Pp being projected into Qq.

The point O is therefore the vertex of a quadric cone which passes through the two conics.

The vertex of another such cone is obtained by producing qP or pQ to meet DF.*

277. *A conic can be projected into a circle so that the projection of any point inside the conic shall be the centre of the circle.*

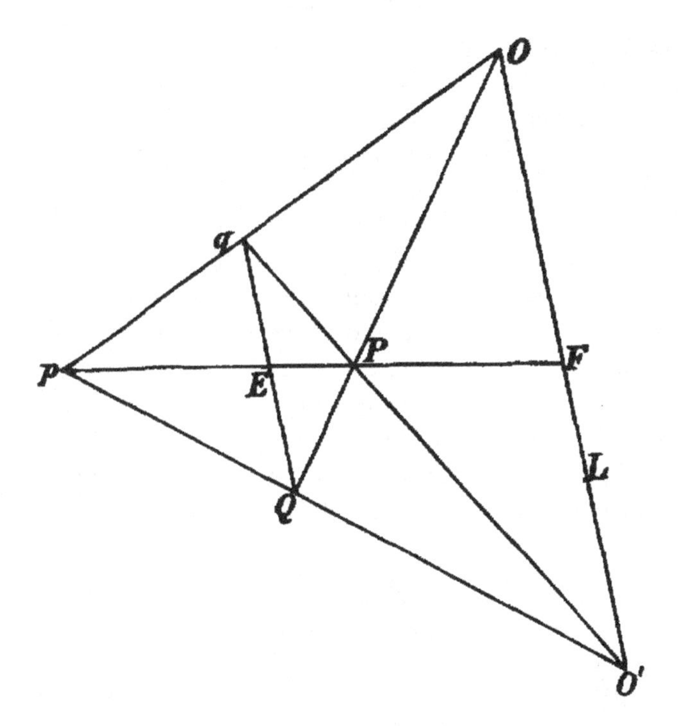

Let E be the point inside the conic and let AB be the chord of which E is the middle point.

*I am indebted to Mr H. F. Baker, Fellow and Lecturer of St John's College, for having called my attention to this theorem and to the mode of proof which is here given. The theorem is given in Poncelet's Treatise, and also in the article on Projections in the last edition of the *Encyclopaedia Britannica*.

Describe a circle on AB as diameter in any plane passing through AB.

Observing that the pole of AEB with regard to the circle is at an infinite distance, draw through F, the pole of AB with regard to the conic, the line FL parallel to that diameter, QEq, of the circle which is perpendicular to AB.

The plane EFL will cut the conic in the diameter Pp, and the circle in the diameter Qq.

If pq, qP intersect FL in O and O', these two points will be vertices from which the conic can be projected into a circle, the centre of which is the projection of the point E.

Since $$FO : Fp :: Eq : Ep :: EA : Ep,$$
it follows that, for different positions of the plane through AB, FO is constant, so that O may be taken anywhere on the circle, centre F, in the plane through F perpendicular to the chord AEB.

Further, $$FO : FP :: EQ : EP :: EA : EP,$$
$$\therefore FO^2 : FP . Fp :: EA^2 : EP . Ep :: CD^2 : CP^2,$$
DCd being the semi-diameter of the conic which is conjugate to CP.

The length FO is thus determined when the position of the point E, inside the conic, is given, and, if we take as the vertex of projection any point O on the circle, centre F, as described above, the projection of the conic on any plane parallel to AEB and FO will be a circle.

If the conic is an ellipse, it follows that FO is equal to the ordinate FR, conjugate to Pp, of the hyperbola in the plane of the ellipse which has the same conjugate diameters PCp and DCd.

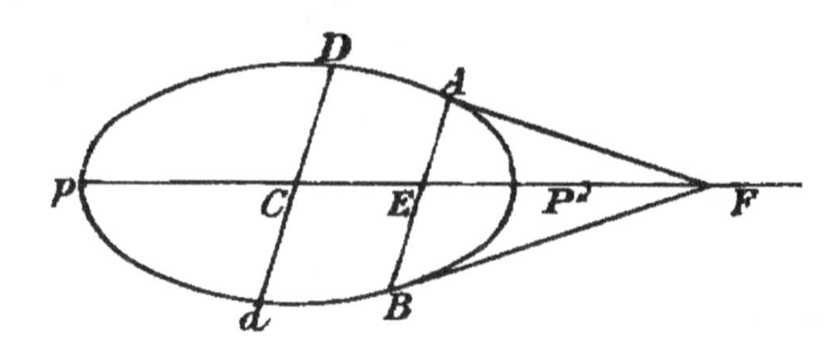

If the conic is an hyperbola, FO is equal to the ordinate FR of an ellipse in the plane of the hyperbola which has the same conjugate diameters PCp and DCd.

This hyperbola or this ellipse constructed outside the given conic may be called *the associated conic.*

If the conic is a parabola, the points O and O' are obtained by drawing lines through q and Q parallel to the axis of the parabola.

In this case,

$$FO^2 = Eq^2 = EA^2 = 4SP\,.\,PE = 4SP\,.\,PF,$$

so that the associated conic is a parabola.

If the conic is a circle, the associated conic is a rectangular hyperbola.

If the conic is an ellipse, the axes of which are indefinitely small, that is, if it is reduced to a point, the associated conic lapses into two straight lines, which are at right angles to each other if the point is the limit of a circle.

278. If the point E be outside the conic, or, in other words, if the polar of E intersect the conic, it is not possible to project the conic into a circle, so that the projection of E shall be the centre of the circle.

In this case the conic can be projected into a rectangular hyperbola, having the projection of the point E for its centre.

Let RU be the chord of contact of the tangents from E, and take any point O on the surface of the sphere of which RU is a diameter.

Then the projection of the conic from the vertex O on any plane parallel to ROU will be an hyperbola, and, since ROU is a right angle, it will be a rectangular hyperbola.

279. *If two conics in a plane are entirely exterior to each other, they can in general be projected, from the same vertex, into circles on the same plane.*

Draw four parallel tangents to the conics, and let F be the point of intersection of the diameters, PCp and QGq, joining the points of contact.

Also, let FR, FR' be the ordinates through F, parallel to the tangents, of the associated conics.

If F is so situated that these ordinates are equal, the locus of the vertices from which the two conics can be projected into circles will be the same, that is, it will be the circle of which F is the centre, and FR the length of the radius, in the plane through F perpendicular to FR.

In this case, taking any point O on the circle as the vertex, the two conics will be projected into circles on any plane parallel to the plane OFR, and the centres of the circles will be the projections of E and E', the respective poles of FR with regard to the conics.

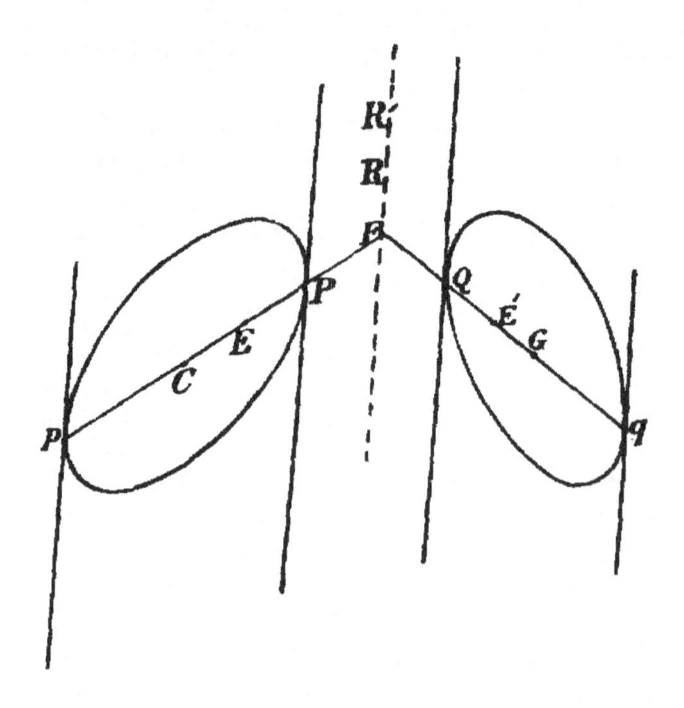

280. For different directions of the tangents, the points, F, R, R', will take up different positions, and for all directions of the tangents the loci of these points will be continuous curves.

The loci of R and R' will, in general, intersect each other; that is to say, there will be, in general, positions of F such that FR and FR' are equal.

Taking a particular case, let F be so situated that FR' is greater than FR; then taking F at the point where its locus meets the conic G, FR' vanishes, and therefore, between these two positions of F, there must be some position such that FR' is equal to FR.

We may observe that the locus of F passes through C and G, the centres of the two conics.

For, if CG is conjugate to the parallel tangents of the conic G, the point F is at C, and, if CG is conjugate to the parallel tangents of the conic C, the point F is at G.

When FR' is equal to FR, the line thus obtained is called by *Poncelet* the *Ideal Secant* of the two conics.

281. In a similar manner if one conic is entirely inside another they can, in general, be projected into circles, one of which will be inside the other.

Also two conics intersecting in two points may be projected into two intersecting circles.

Two conics intersecting in four points, or having contact at two points, cannot be projected into circles, but they can be projected into rectangular hyperbolas.

282. The method of projections enables us to extend to conics theorems which have been proved for a circle, and which involve, amongst other ideas, harmonic ranges, poles and polars, systems of collinear points, and systems of concurrent lines.

For instance, the theorems of Arts. 208 and 210 are easily proved for a circle, and by this method are at once extended to conics.

Take as another instance Pascal's theorem, that *the opposite sides of any hexagon inscribed in a conic intersect in three collinear points.*

If this be proved for a circle, the method of conical projection at once shews that it is true for any conic.

The following very elementary proof of the theorem for a circle is given in *Catalan's Théorèmes et Problèmes de Géométrie Elémentaire.*

Let $ABCDEF$ be the hexagon, and let AB and ED meet in G, BC and FE in H, FA and DC in K.

Also let ED meet BC in M and AF in N, and let BC meet AF in L.

Then we have the relations,

$$LA \,.\, LF = LB \,.\, LC, MC \,.\, MB = MD \,.\, ME,$$

$$NE \,.\, ND = NF \,.\, NA.$$

Also, the triangle LMN being cut by the three transversals AG, DK, FH, we have the relations,

$$LB \,.\, MG \,.\, NA = LA \,.\, MB \,.\, NG$$

$$LC \,.\, MD \,.\, NK = LK \,.\, MC \,.\, ND$$

$$LH \,.\, ME \,.\, NF = LF \,.\, MH \,.\, NE.$$

Multiplying together these six equalities, taking account of the relations previously stated, and cutting out the factors common to the two products, we obtain

$$LH \,.\, MG \,.\, NK = LK \,.\, MH \,.\, NG;$$

$\therefore G$, H, K are collinear.

Brianchon's theorem that, *if a hexagon circumscribe a conic, the three opposite diagonals are concurrent* is proved at once by observing that it is the reciprocal polar of Pascal's theorem.

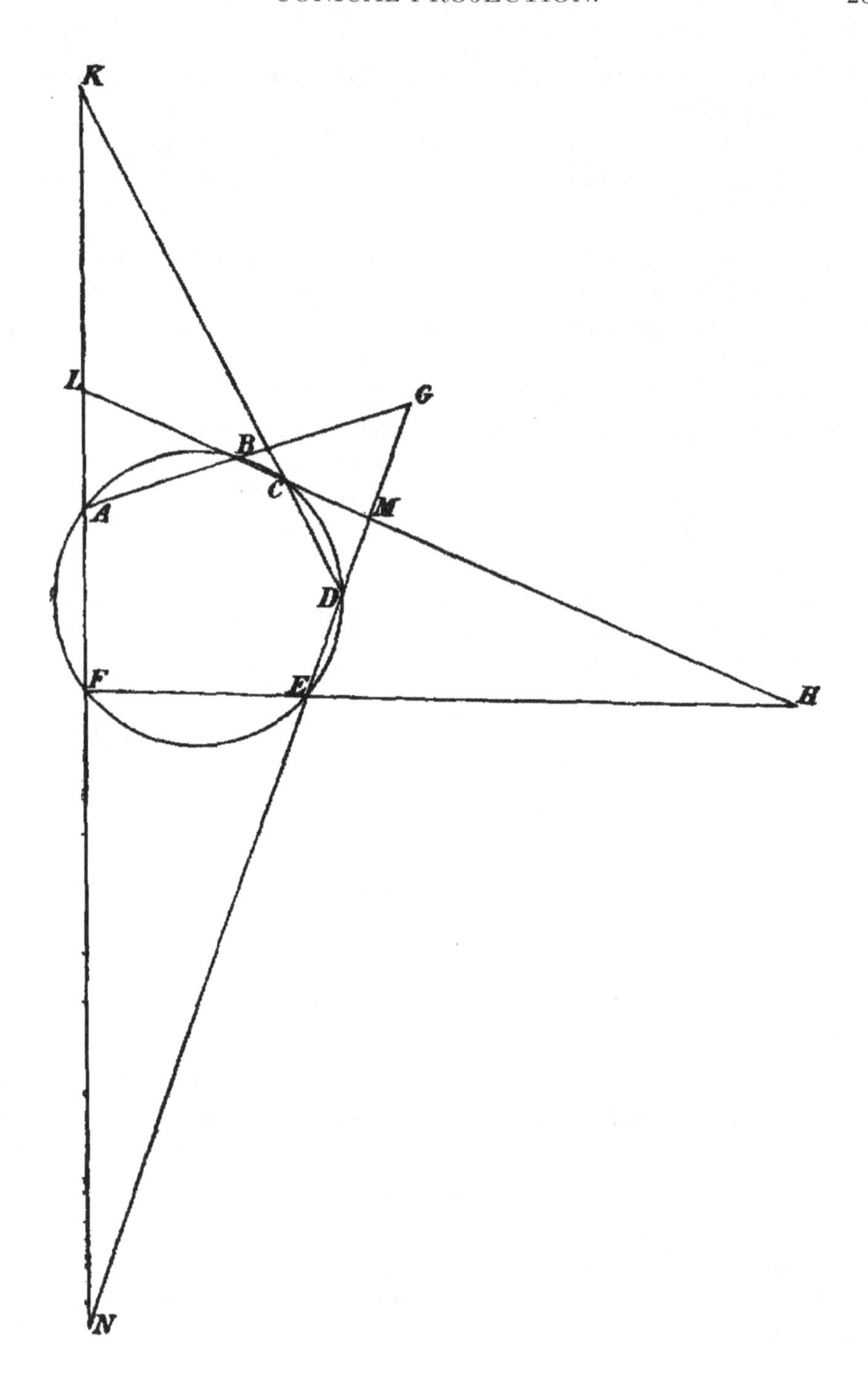
K
L
G
B
C
A
M
D
F
E
H
N

283. *Stereographic and Gnomonic Projections.*

If a point on the surface of a sphere be taken as the vertex of projection, and if the plane of projection be parallel to the tangent plane at the point, the projection of any figure drawn on the surface of the sphere is called its stereographic projection.

If however the centre of the sphere be taken as the vertex of projection, and any plane be taken as the plane of projection, the projection of any figure drawn on the surface of the sphere is called its gnomonic projection.

The stereographic projection of a circle drawn on the surface of the sphere is a circle; for it can be easily shewn that it is a subcontrary section of the oblique cone formed by the vertex of projection and the circle on the sphere.

The gnomonic projection of a circle on the sphere is obviously a conic.

These projections are sometimes described in treatises on Astronomy, and in these treatises the vertex for stereographic projection is taken at the south pole of the earth, and, for gnomonic projection, at the centre of the earth; and, in both cases, the plane of projection is taken parallel to the plane of the equator.

284. It will be seen that the discussions which are given in this chapter are confined entirely to cases of *real* projection.

The chapter is intended to be simply an introduction to a large and important subject.

The method of conical projections is due to Poncelet, and is worked out with great fulness and elaboration in his work entitled, *Traité des Propriétés Projectives des Figures* (Second edition, 1865, in two quarto volumes).

In this work Poncelet extends the domain of pure geometry by the interpretation and use of the law of continuity, and, as one of its applications, by the introduction of the imaginary chord of intersection, or, as it is called by Poncelet, the ideal secant of two conics.

Amongst English writers, the student will find valuable chapters on projections in *Salmon's Conics*, and in the large work on the *Geometry of Conics, by Dr C. Taylor, the Master of St John's College, Cambridge.*

There is also an important work by *Cremona, on Projective Geometry*, which has been translated by *Leudesdorf* (Second edition, 1893).

MISCELLANEOUS PROBLEMS. II.

1. If two conics have the same directrix, their common points are concyclic.

2. If a focal chord of a parabola is bisected in V and the line perpendicular to it through V meets the axis in G, SG is half the chord.

3. If the perpendicular to CP from a point P of an ellipse meets the auxiliary circle in Q, PQ varies as PN.

4. AA' and BB' are the axes, and S is one of the foci of an ellipse; if a parabola is described with S as focus and passing through B and B', its vertex bisects SA or SA'.

5. Tangents to an ellipse at P, p intersect on an axis; if the perpendicular from p on the tangent at P intersects CP in L, the locus of L is a similar ellipse.

6. The normal to a hyperbola at P meets the axes in G and g respectively. Prove that the circle circumscribing SPG is touched Sg.

7. If a tangent to an ellipse meets a pair of conjugate diameters in points equidistant from the centre, the locus of the points is a circle.

8. If ellipses are described on AB as diameter, touching BC, the points of contact of tangents from C are on a straight line.

9. If Pl, Pm be drawn perpendicular to CL, CM respectively, shew that the centre of the circle Plm lies on a fixed hyperbola.

10. PSQ, PHR are focal chords of an ellipse, QT, RT the tangents at Q and R. Shew that PT is the normal at P.

11. A, B are two fixed points. Through them a system of circles is drawn. Through A draw any two lines meeting the circles in the points C_1D_1, C_2D_2, &c. Shew that the lines CD all touch a parabola, focus B, which also touches the lines AC, AD.

12. From any two points A, B on an ellipse four lines are drawn to the foci S, H. Shew that $SA \,.\, HB$ and $SB \,.\, HA$ are to one another as the squares of the perpendiculars from a focus on the tangents at A and B.

13. If two points of a conic and the angle subtended by these points at the focus are given, the line joining the focus with the intersection of the tangents always passes through a fixed point.

14. If normals to an ellipse are drawn at the extremities of chords parallel to one of the equi-conjugate diameters, pairs of such normals intersect on the line through the centre perpendicular to the other diameter.

15. From the point in which the tangent at any point P of a hyperbola cuts either asymptote perpendiculars are dropped upon the axes. Prove that the line joining the feet of these perpendiculars passes through P.

16. Tangents are drawn to an ellipse parallel to conjugate diameters of a second given ellipse. Shew that the locus of their intersection is an ellipse similar and similarly situated to the second ellipse.

17. A focus of a conic inscribed in a triangle being given, find the points of contact.

18. The normals at P and Q, the ends of a focal chord PSQ, intersect in K, and KN is perpendicular to PQ; prove that NP and SQ are equal.

19. If CR, SY, HZ be perpendiculars upon the tangent at a point P such that $CR = CS$, prove that R lies on the tangent at B, and that the perpendicular from R on SH will divide it into two parts equal to SY, HZ respectively.

20. If a parabola, having its focus coincident with one of the foci of an ellipse, touches the conjugate axis of the ellipse, a common tangent to the ellipse and parabola will subtend a right angle at the focus.

21. Two tangents TP and TQ are drawn to an ellipse, and any chord TRS is drawn, V being the middle point of the intercepted part; QV meets the ellipse in P'; prove that PP' is parallel to ST.

22. If S, S' are the foci of an ellipse and SY, $S'Y'$ the perpendiculars on any tangent, XY, $X'Y'$ meet on the minor axis, and, if PN is the ordinate of P, NY and NY' are perpendicular to XY and $X'Y'$ respectively.

23. A circle through the centre of a rectangular hyperbola cuts the curve in the points A, B, C, D. Prove that the circle circumscribing the triangle formed by the tangents at A, B, C passes through the centre of the hyperbola.

24. If the tangent at a point P of an ellipse meets any pair of parallel tangents in M, N, and if the circle on MN as diameter meets the normal at P in K, L,

then KL is equal to DCD', and CK, CL are equal to the sum and difference of the semi-axes.

25. From a point O two tangents OA, OB are drawn to a parabola meeting any diameter in P, Q. Prove that the lines OP, OQ are similarly divided by the points of contact, but one internally, the other externally.

26. If S, H be the foci of an ellipse, and SP, HQ be parallel radii vectores drawn towards the same parts, prove that the tangents to the ellipse at P, Q intersect on a fixed circle.

27. If an ellipse be inscribed in a quadrilateral so that one focus S is equidistant from the four vertices, the other focus must be at the intersection H of the diagonals.

28. P is a point on a circle whose centre is Q; through P a series of rectangular hyperbolas are described having Q for their centre of curvature at P. Prove that the locus of their centres is a circle with diameter of length PQ.

29. Two cones which have a common vertex, their axes at right angles, and their vertical angles supplementary, are intersected by a plane at right angles to the plane of their axes. Prove that the distances of either focus of the elliptic section from the foci of the hyperbolic section are equal respectively to the distance from the vertex of the ends of the transverse axis of each, and that the sum of the squares on the semi-conjugate axes is equal to the rectangle contained by those distances.

30. Two plane sections of a cone which are not parallel are such that a focus of each and the vertex of the cone lie on a straight line. Shew that the angle included by any pair of focal chords of one section is equal to that contained by the corresponding focal chords of the other section, corresponding chords being the projections of each other with respect to the vertex.

31. If PP', QQ' be chords normal to a conic at P and Q, and also at right angles to each other, then will PQ be parallel to $P'Q'$.

32. A system of conics have a common focus S and a common directrix corresponding to S. A fixed straight line through S intersects the conics, and at the points of intersection normals are drawn. Prove that these normals are all tangents to a parabola.

33. If two confocal conics intersect, prove that the centre of curvature of either curve at a point of intersection is the pole of the tangent at that point with regard to the other curve.

34. A chord of a conic whose pole is O meets the directrices in R and R'; if SR and HR' meet in O', prove that the minor axis bisects OO'.

35. TQ and TR, tangents to a parabola, meet the tangent at P in X and Y, and TU is drawn parallel to the axis, meeting the parabola in U. Prove that the tangent at U passes through the middle point of XY, and that, if S is the focus,

$$XY^2 = 4SP\,.\,TU.$$

36. The foot of the directrix which corresponds to S is X, and XY meets the minor axis in T; CV is the perpendicular from the centre on the tangent at P. Prove that, if $CP = CS$, then $CV = VT$.

37. A is a given point in the plane of a given circle, and ABC a given angle. If B moves round the circumference of the circle, prove that, for different values of the angle ABC, the envelopes of BC are similar conics, and that all their directrices pass through one or other of two fixed points.

38. If AA' is the transverse axis of an ellipse, and if Y, Y' are the feet of the perpendiculars let fall from the foci on the tangent at any point of the curve, prove that the locus of the point of intersection of AY and $A'Y'$ is an ellipse.

39. The tangent at a point P of an hyperbola cuts the asymptotes in L and L', and another hyperbola having the same asymptotes bisects PL and PL'. Prove that it intersects CP in a point p such that

$$Cp^2 : CP^2 :: 3 : 4.$$

The chord QR, joining a point R on an asymptote with a point Q on the corresponding branch of the first hyperbola, intersects the second hyperbola in E; if QR move off parallel to itself to infinity, prove that, ultimately $RE : EQ :: 3 : 1$.

40. Tangents are drawn to a rectangular hyperbola from a point T in the transverse axis, meeting the tangents at the vertices in Q and Q'. Prove that QQ' touches the auxiliary circle at a point R such that RT bisects the angle QTQ'.

41. Tangents from a point T touch the curve at P and Q; if PQ meet the directrices in R and R', PR and QR' subtend equal angles at T.

42. The straight lines joining any point to the intersections of its polar with the directrices touch a conic confocal with the given one.

43. If a point moves in a plane so that the sum or difference of its distances from two fixed points, one in the given plane and the other external to it, is constant, it will describe a conic, the section of a right cone whose vertex is the given external point.

44. In the construction of Art. 241 prove that CK' and CK are respectively equal to the sum and difference of the semi-axes.

45. Given a tangent to an ellipse, its point of contact, and the director circle, construct the ellipse.

46. If the tangent at any point P of an ellipse meet the auxiliary circle in Q', R', and if Q, R be the corresponding points on the ellipse, the tangents at Q and R pass through the point P' on the auxiliary circle corresponding to P.

47. In the ellipse $PDP'D'$, $P'HCSPX$ and DCD' are conjugate diameters; CH is equal to CS, and the polar of S passes through a point X on $P'P$ produced. If DX is drawn cutting the ellipse in Q, prove that HD is parallel to SQ.

48. If T is the pole of a chord of a conic, and F the intersection of the chord with the directrix, TSF is a right angle.

49. The polar of the middle point of a normal chord of a parabola meets the focal vector to the point of intersection of the chord with the directrix on the normal at the further end of the chord.

50. OP, OQ touch a parabola at P, Q; the tangent at R meets OP, OQ in S, T; if V is the intersection of PT, SQ, O, R, V are collinear.

51. If from any point A a straight line AEK be drawn parallel to an asymptote of an hyperbola, and meeting the polar of A in K and the curve in E, shew that $AE = EK$.

52. If a chord PQ of a parabola, whose pole is T, cut the directrix in F, the tangents from F bisect the angle PFT and its supplement.

53. A parabola, focus S, touches the three sides of a triangle ABC, bisecting the base BC in D; prove that AS is a fourth proportional to AD, AB, and AC.

54. A focal chord PSQ is drawn to a conic of which C is the centre; the tangents and normals at P and Q intersect in T and K respectively; shew that ST, SP, SK, SC form an harmonic pencil.

55. PCP' is any diameter of an ellipse. The tangents at any two points D and E intersect in F. PE, $P'D$ intersect in G. Shew that FG is parallel to the diameter conjugate to PCP'.

56. A conic section is circumscribed by a quadrilateral $ABCD$: A is joined to the points of contact of CB, CD; and C to the points of contact of AB, AD; prove that BD is a diagonal of the interior quadrilateral thus formed.

57. A parabola touches the three lines CB, CA, AB in P, Q, R, and through R a line parallel to the axis meets RQ in E; shew that $ABEC$ is a parallelogram.

58. If a series of conics be inscribed in a given quadrilateral, shew that their centres lie on a fixed straight line.

Shew also that this line passes through the middle points of the diagonals.

59. Four points A, B, C, D are taken, no three of which lie in a straight line, and joined in every possible way; and with another point as focus four conics are described touching respectively the sides of the triangles BCD, CDA, DAB, ABC; prove that the four conics have a common tangent.

60. If the diagonals of a quadrilateral circumscribing a conic intersect in a focus, they are at right angles to one another, and the third diagonal is the corresponding directrix.

61. An ellipse and parabola have the same focus and directrix; tangents are drawn to the ellipse at the extremities of the major axis; shew that the diagonals of the quadrilateral formed by the four points where these tangents cut the parabola intersect in the common focus, and pass through the extremities of the minor axis of the ellipse.

62. Three chords of a circle pass through a point on the circumference; with this point as focus and the chords as axes three parabolas are described whose parameters are inversely proportional to the chords; prove that the common tangents to the parabolas, taken two and two, meet in a point.

63. A circle is described touching the asymptotes of an hyperbola and having its centre at the focus. A tangent to this circle cuts the directrix in F, and has its pole with regard to the hyperbola at T. Prove that TF touches the circle.

64. Two conics have a common focus: their corresponding directrices will intersect on their common chord, at a point whose focal distance is at right angles to that of the intersection of their common tangents. Also the parts into which either

common tangent is divided by their common chord will subtend equal angles at the common focus.

If the conics are parabolas, the inclination of their axes will be the angle subtended by the common tangent at the common focus.

65. The tangent at the point P of an hyperbola meets the directrix in Q; another point R is taken on the directrix such that QR subtends at the focus an angle equal to that between the transverse axis and an asymptote; prove that the envelope of RP is a parabola.

66. If an hyperbola passes through the angular points of an equilateral triangle and has the centre of the circumscribing circle as focus, its eccentricity is the ratio of 4 to 3, and its latus rectum is one-third of the diameter of the circle.

67. An isosceles triangle is circumscribed to a parabola; prove that the three sides and the three chords of contact intersect the directrix in five points, such that the distance between any two successive points subtends the same angle at the focus.

68. Tangents are drawn at two points P, P' on an ellipse. If any tangent be drawn meeting those at P, P' in R, R', shew that the line bisecting the angle RSR' intersects RR' on a fixed tangent to the ellipse.

69. The chords of a conic which subtend the same angle at the focus all touch another conic having the same focus and directrix.

70. Two conics have a common focus S and a common directrix, and tangents TP, TP' are drawn to one from any point on the other and meet the directrix in F and F'. Prove that the angles PSF', $P'SF$ are equal and constant.

71. A rectangular hyperbola circumscribes a triangle ABC; if D, E, F are the feet of the perpendiculars from A, B, C on the opposite sides, the loci of the poles of the sides of the triangle ABC are the lines EF, FD, DE.

72. If two of the sides of a triangle, inscribed in a conic, pass through fixed points, the envelope of the third side is a conic.

73. If two circles be inscribed in a conic, and tangents be drawn to the circles from any point in the conic, the sum or difference of these tangents is constant, according as the point does or does not lie between the two chords of contact.

74. The four common tangents of two conics intersect two and two on the sides of the common self-conjugate triangle of the conics.

75. Prove that a right cylinder, upon a given elliptic base, can be cut in two ways so that the curve of section may be a circle; and that a sphere can always be drawn through any two circular sections of opposite systems.

76. An ellipse revolves about its major axis, and planes are drawn through a focus cutting the surface thus formed. Prove that the locus of the centres of the different sections is a surface formed by the revolution of an ellipse about CS where C or S are respectively the centre and focus of the original ellipse.

77. Given five tangents to a conic, find, by aid of Brianchon's theorem, the points of contact.

78. The alternate angular points of any pentagon $ABCDE$ are joined, thus forming another pentagon whose corresponding angular points are a, b, c, d, e; Aa, Bb, Cc, Dd, Ee are joined and produced to meet the opposite sides of $ABCDE$ in α, β, γ, ϵ; shew that if A be joined with the middle point of $\gamma\delta$, B with the middle point of $\delta\epsilon$, &c., these five lines meet in a point.

79. If a conic be inscribed in a triangle, the lines joining the angular points to the points of contact of the opposite sides are concurrent.

80. If a quadrilateral circumscribe a conic, the intersection of the lines joining opposite points of contact is the same as the intersection of the diagonals.

81. ABC is a triangle, and D, E, F the middle points of the sides. Shew that any two similar and similarly situated ellipses one circumscribing DEF and the other inscribed in ABC will touch each other.

82. AB is a chord of a conic. The tangents at A and B meet in T. Through B a straight line is drawn meeting the conic in C and AT in P. The tangent to the conic at C meets AT in Q. Prove that $TPQA$ is a harmonic range.

83. Pp, Qq, Rr, Ss are four concurrent chords of a conic; shew that a conic can be drawn touching SR, RQ, QP, sr, rq, qp.

84. If two sections of a right cone have a common directrix, the latera recta are in the ratio of the eccentricities.

85. $ABCD$ is a parallelogram and a conic is described to touch its four sides. If S is a focus of this conic and if with S as focus a parabola is described to touch AB and BC, the axis of the parabola passes through D.

86. If from a point O tangents be drawn to two conics S and S', and if the tangents to S be conjugate with respect to S', prove that the tangents to S' are conjugate with respect to S.

87. If a triangle is self-conjugate with respect to each of a series of parabolas, the lines joining the middle points of its sides will be tangents; all the directrices will pass through O, the centre of the circumscribing circle; and the focal chords, which are the polars of O, will all touch an ellipse inscribed in the given triangle which has the nine-point circle for its auxiliary circle.

88. If a triangle can be drawn so as to be inscribed in one given conic and circumscribed about another given conic, an infinite number of such triangles can be drawn.

89. Prove that the stereographic projection of a series of parallel circles on a sphere is a series of coaxal circles, the limiting points of which are the projections of the poles of the circles.

90. Through the six points of intersection of a conic with the sides of a triangle straight lines are drawn to the opposite angular points; if three of these lines are concurrent the other three are also concurrent.

91. Prove that the asymptotes of an hyperbola, and a pair of conjugate diameters form an harmonic range, and that the system of pairs of conjugate diameters is a pencil in involution.

92. If two concentric conics have the directions of two pairs of conjugate diameters the same, then the directions are the same for every pair.

93. If two concentric conics have all pairs of conjugate diameters in the same directions, and have a common point, they coincide entirely.

94. If two conics have two common self-conjugate triangles with the same vertex, which is interior to both, they cannot intersect in any point without entirely coinciding.

95. If two conics in space whose planes intersect in a line which does not cut either conic, and if on this line there are four points, P, P', Q, Q', such that the polars of P with regard to the conics both pass through P', and that the polars of Q both pass through Q', then either conic can be projected into the other in two ways.

CAMBRIDGE: PRINTED BY J. AND C. F. CLAY, AT THE UNIVERSITY PRESS.

Made in the USA
Las Vegas, NV
14 October 2023